SELECTED WORKS OF MODERN CHINESE LEARNING

KEY ECONOMIC AREAS IN CHINESE HISTORY

Ch'ao-ting Chi

2020 • BEIJING

First Edition 2014

ISBN 978 - 7 - 100 - 10501 - 9

Published by

The Commercial Press

36 Wangfujing Road, Beijing 100710, China

www.cp.com.cn

Ch'ao-Ting Chi, Ph.D.

(1903—1963)

Editorial Note

One hundred years ago, Zhang Zhidong tried to advocate Chinese learning by saying: "The course of a nation, be it bright or gloomy, the pool of talents, be it large or small, are about governance on the surface, and about learning at the root." At that time, the imperialist powers cast menacing eyes on our country, and the domestic situation was deteriorating. The quick infiltration of Western learning made the long-standing Chinese tradition come under heavy challenge. In those days, Chinese learning and Western learning stood side by side. Literature, history and philosophy split up, while many new branches of learning such as economics, politics and sociology were flourishing, which made many Chinese dazed. However, there appeared a vital and vigorous learning climate out of the confusing situation. It was at this critical moment that modern Chinese scholarship made the transition—by exchanging views, basing on profound contemplation and even with confrontation of idea and clash of views, the scholarship made continuous progress, bringing up a large number of persons of academic distinction and creating numerous innovative works. Changes in scholarship and in general modes of thinking made transition in all aspects of the society possible, thus laying a solid foundation for revitalizing China.

It's over a century since the journey of modern Chinese learning started, during which various schools of thought stood in great numbers, causing heated discussions. The journey sees schools of thought as well as relevant arguments rising and falling, waxing and waning instantly, leaving complicated puzzles to followers. By studying and reviewing the selected works, one may gain new insights into that journey; and it is the editor's sincere hope that readers would ponder over the future by recalling the past. That's why we have compiled "Selected Works of Modern Chinese Learning". The effort includes masterpieces of celebrated scholars from diverse fields of study and different schools of thought. By tracing back to the source and searching for the basis of modern Chinese learning, we wish to present the dynamics between thought and time.

The series of "Selected Works of Modern Chinese Learning" includes works (both in Chinese and in foreign languages) of scholars from China—mainland, Hong Kong, Macau, and Taiwan—and from overseas. These works are mostly on humanities and cover all fields of subjects, such as literary theory, linguistics, history, philosophy, politics, economics, jurisprudence, sociology, to name a few.

It has been a long-cherished wish of the Commercial Press to compile a series of "Selected Works of Modern Chinese Learning". Since its foundation in 1897, the Commercial Press has been privileged to have published numerous pioneering works and masterpieces of modern Chinese learning under the motto of "promoting education and enlightening people". The press has participated in and witnessed the establishment and development of modern Chinese learning. The series of "Selected Works of Modern Chinese Learning" is fruit of an effort to relay the editorial legacy and the cultural propositions of our senior generations. This series, sponsored by National Publication Foundation, would not be possible if there were no careful planning of the press itself. Neither would it be possible without extensive collaboration among talents of the academic circle. It is our deeply cherished hope that titles of this series will keep their place on the bookshelves even after a long time. Moreover, we wish that this series and "Chinese Translations of World Classics" will become double jade in Chinese publishing history as well as in the history of the Commercial Press itself. With such great aspirations in mind, fearing that it is beyond our ability to realize them, we cordially invite both scholars and readers to extend your assistance.

Editorial Department of the Commercial Press
December 2010

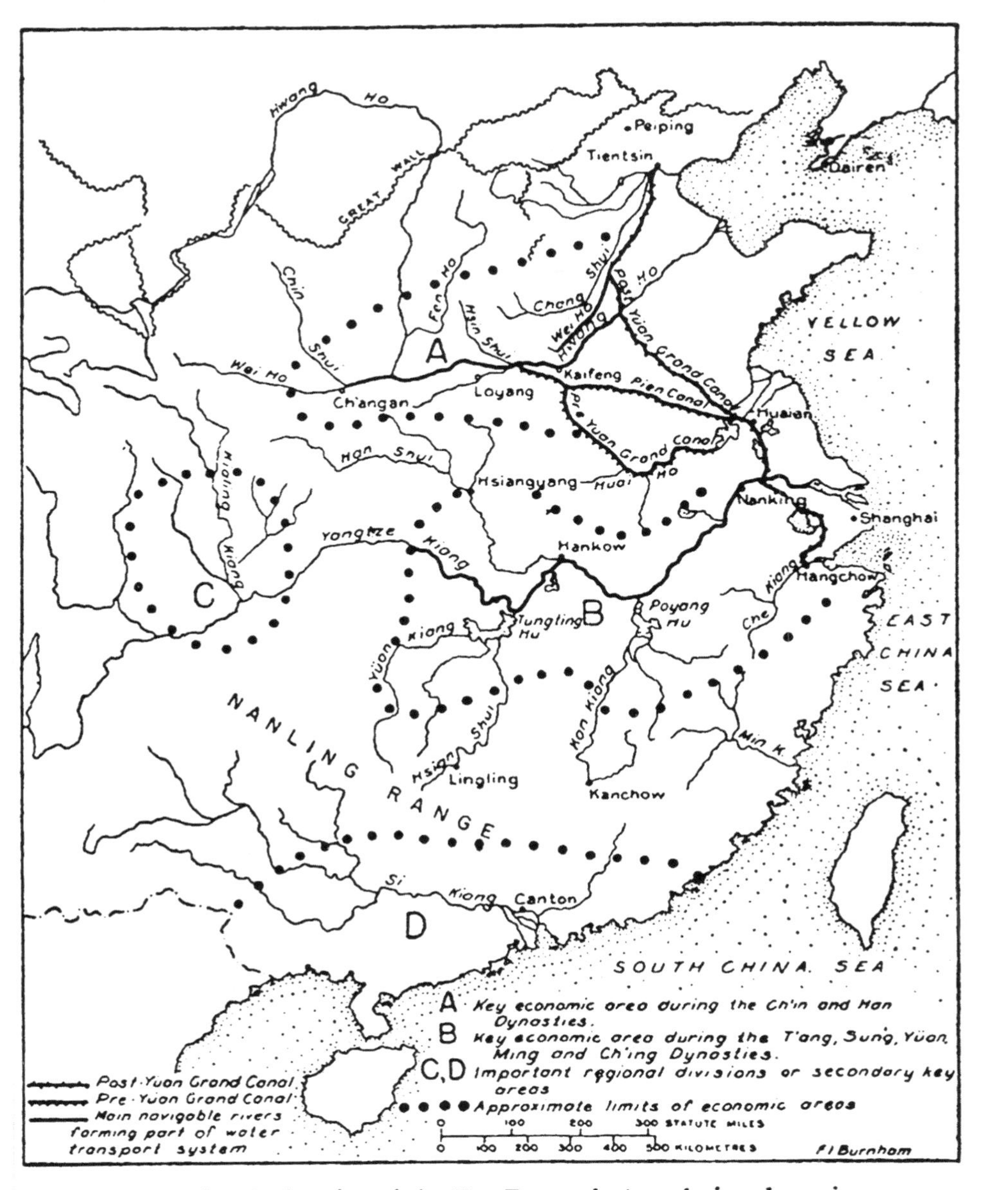

Map showing the location of the Key Economic Area during the various periods of Chinese History

TO

HARRIET

WITH

LOVE, RESPECT AND GRATITUDE

PREFACE

THE present work offers the conception of the dynamics of the Key Economic Area as an aid to the understanding of Chinese economic history. By tracing the development of the Key Economic Area through an historical study of the construction of irrigation and flood-control works and transport canals, it aims to show the function of the Key Economic Area as an instrument of control of subordinate areas and as a weapon of political struggle, to indicate how it shifts, to reveal its dynamic relation to the problem of unity and division in Chinese history, and to give, on the basis of this approach, a concrete and historical-descriptive analysis of one phase of the economic development of China. The book does not purport to give a new interpretation of Chinese history as a whole. However, if the concept of the Key Economic Area proves helpful to the solution of one of the fundamental problems in Chinese history, it cannot but affect the understanding and interpretation of the whole process of Chinese historical development. In order to place this theory in its proper perspective and to indicate its possible extension and further application, it is perhaps better at this point for the author to state his general approach to historical economic studies in China.

Man makes his own history, not only under conditions which history hands down to him, but also through the rewriting of past history. This is because history itself is historical and can only be understood by each epoch, and be of service to it, in the light of its own experience. New experience gives rise to new historical insight, and in the light of new understanding, new problems can be formulated, old and new evidence resifted, and significant facts selected out of a multitude of seemingly meaningless data. Thus history must be continually rewritten in order to answer the need of man in each specific epoch. The rewriting of history is part of man's efforts to harness the forces of history, and this task becomes particularly urgent at each turning point in the historical process.

Ever since the "opening of China" in the middle of the nineteenth century, the problem of rewriting Chinese history has loomed large on the intellectual horizon not only of China, but of the whole world. Capitalist foreign trade and investment, through its exploitation of the world market, has created an interdependent world economy and has thrown the unevenly and differently developed socio-economic set-ups of various peoples into one turbulent current of world history. Chinese history is no longer the history of just one country, but it has merged into the stream of world history. "Western" institutions have made serious inroads into Chinese life, which, in turn, has become an important factor in the life of the "West." The far-reaching consequences of the situation were epitomized in the Great Revolution of 1925-1927 and its subsequent developments, which brought to the fore, for the first time in centuries, the most fundamental problems of the dynamics of Chinese society.

Economic history, or dialectical economics, recognizes the fact that fundamental problems of China to-day cannot be understood merely by studying contemporary conditions, but must be approached historically through attempts to solve the basic questions of Chinese history raised by the demands of our epoch, and to discover the dominant tendencies which govern the development of Chinese economy. The main objective of such a study, of course, is historical and socio-economic synthesis.

But synthesis and analysis are two phases of the same process and cannot be mechanically separated. Synthesis signifies building up or organization, while analysis means breaking down or separation. But, since both exclude chaos and wanton behaviour, it is clear that building up is impossible without first having broken down and understood the meaning of the parts, and it is equally true that taking apart is impossible without first having an idea of how the parts are put together. Applying this principle to historical writing, it means that synthesis involves the systematic merging of leading ideas which result from analytical study of special problems, while analysis cannot be fruitful without a general approach to guide its labours in working through a maze of otherwise meaningless data. The apparent contradiction between the two concepts is really a reflection of their dialectical relationship; both represent necessary phases of the same process of scientific investigation. A book may be primarily a work of synthesis or of analysis, but an investigation can be fruitful only when the intimate connection between the two concepts is expressly or tacitly recognized.

Although the final purpose of the author's study of economic history is synthesis, the present book is primarily a work of analysis. It began as an attempt to trace the development of irrigation and flood-control in Chinese history through an analytical study of the immense amount of untouched source material on the subject hidden in the gazetteers (local historical geographies), special Chinese works on "water benefits," and the dynastic histories. The general direction of the author's researches naturally has been determined by his preconceptions and general method of approach, specifically by his realization of the importance of irrigation and water transportation in Chinese history. But it was only after a close examination of most of the available material in the Library of Congress, Washington, D.C., that the belief in the importance of water-control to Chinese history was confirmed and the conception of the Key Economic Area and its relation to unity and division in Chinese history developed in the author's mind. This book as it stands, which, except for minor changes, was completed in the middle of April 1934, represents the author's preliminary effort to define the concept, to study the geographical basis for water-control and economic regionalization in China, and to trace briefly the shifting of the Key Economic Area in Chinese history.

Thus the analytical study of source material on the history of water-control gave birth to a new concept which has in turn been used as a device to define the course of development of water-control. But a concept is like a lamp—once lit it does not illuminate merely one corner of the room. The concept of the Key Economic Area throws light on every fundamental problem in Chinese history. It emphasizes the local and regional character of Chinese economy. The First Emperor, Ch'in Shih Huang Ti (221-210 B.C.), successfully battered down the feudal barriers between the warring states, but the unity he achieved was a loose unity. It was not bound together by economic ties like those in a modern state, but was held together by military and bureaucratic domination through the instrumentality of the control of the Key Economic Area. Such a unity cannot be enduring, and as soon as the supremacy of the Key Economic Area was challenged, the ruling power lost its place of mooring and source of supply. Division and chaos then became the order of the day until a new power rooted itself in a Key Economic Area and successfully used it as a weapon for re-establishing unity. This is the true meaning behind the classical Chinese saying that "A long period of unity must eventuate in division; a long period of division must eventuate in unity." It is an iron law describing accurately one of the fundamental movements in the semi-feudal phase of China's history, from the first Emperor to the breakdown of China's isolation in the last century.

The existence of the Key Economic Area motivated the geographical differentiation in the land system and methods of taxation, and accentuated the natural tendency toward uneven development of the different regions. It also affected the distribution of merchant capital and created varied conditions for its development. Differences in the land system, taxation and the degree of development of commercial and usurious capital meant differences in the social characteristics and power of the local ruling groups, differences in the degree of exploitation, and differences in the conditions of the life and work of the peasants, the overwhelming majority of the population. While these differences were never of such a nature as to alter the picture of an essentially homogeneous structure of society in the wide territory of China, they were sufficiently important to influence materially the development of the multifarious phases of the class struggle, especially the peasant wars. Furthermore, although the theory of the Key Economic Area does not explain the persistent tendency to latifundia, the growth of merchant capital and forces retarding its development, the antagonisms between the bureaucracy, the landowners, the merchants, and the peasants, and the periodic decline and breakdown of internal economy—in a word, although this theory does not explain the class struggle, it does reveal one of the important influences shaping the course of development of that struggle.

Let us consider for a moment the influence of the Key Economic Area on the peasant wars, which have always been the culmination and sharpest expression of China's social crises, and, therefore, deserve to be studied as the corner-stones of Chinese history. When a socio-economic cycle, which usually coincided with a dynastic period, drew to a close, when exploitation of the peasants increased and production declined, when extravagance and corruption weakened the ruling power, and when bankruptcy faced the government and starvation confronted the pauperized population, the peasants usually took the road of rebellion, refusing to pay rent, taxes and debts, harassing and expropriating the rich and sacking centres of political power and administration. Such a movement usually started with a series of rebellions in various localities. Owing to the locally self-sufficient economy of the country, consolidation of the scattered forces of the peasants was difficult and speedy conclusion of any such struggle was well-nigh impossible. In the course of the protracted struggle, geographical and economic conditions of the localities limited the growth of certain groups and favoured the development of others. The struggle was one against the government as well as a painful process of elimination among rival groups. Usually the group most favoured by geographical and economic advantages and able leadership emerged from the struggle as the master of the situation. The importance of the geographical locus is considered here not so much from the point of view of strategy as from the point of view of economics. When the problem is studied concretely, in a history of the peasant wars, which the author plans to do, it will be seen that the theory of the Key Economic Area will help an understanding of many hitherto unexplained or misunderstood events.

Thus, despite the fact that the question of regional relations is not the central question in Chinese history and the concept of the Key Economic Area cannot be the ruling idea in the interpretation of Chinese history as a whole, it is important to realize that under conditions of regional natural economy the theory has a special significance, advances considerably our understanding of the whole process of Chinese history, and provides a background for a study of the effects of the impact of imperialism and the problems of contemporary China. Before the full import of the idea can be realized, many articles and monographs dealing with specific phases of its application will have to be written. On the question of the history of water-control alone, the author has collected much more material than he has been able to use. The present volume can only be considered a beginning. Its ruling concept remains to be tested, to be related to other prevailing tendencies in Chinese history, and its manifold implications as outlined in the two foregoing paragraphs still have to be worked out. On the basis of these possibilities the author is planning new projects of research. If circumstances permit, he hopes that the rich historical literature in China, most of which has not yet been scientifically examined, will give rise to other helpful ideas that may confirm, reinforce, or modify the concept of the Key Economic Area.

The author wishes to express his sincere appreciation to the staff of the Library of Congress, Washington, D.C., particularly Dr Arthur W. Hummel, Chief of the Division of Orientalia, and his assistants Mr B. A. Claytor, Dr Shio Sakanishi, and Mr Shao-hsüan Han, for making available for his use the splendid collection of the Library and providing him with practically ideal physical conditions for research. He is deeply indebted to Professors Vladimir G. Simkhovitch, John E. Orchard, and L. C. Goodrich for their keen interest in his work and the reading and criticizing of the whole manuscript. He is also obliged to Dr K. A. Wittfogel whose illuminating contributions in Chinese economic and social history have proved valuable guiding posts to many other workers in the field, and who has kindly read the whole manuscript and offered many valuable suggestions. The author's gratitude is also due to his friends, Mrs Rose Marcus Coe for reading the proofs of the manuscript and Mr Frank Virginia Coe for reading the manuscript, and to both for giving valuable criticism. To Mr Owen Lattimore, Editor of *Pacific Affairs*, he owes a debt that he cannot expect to repay. Mr Lattimore painstakingly went through the whole manuscript, suggested countless important and detailed editorial corrections, and kindly recommended the book for publication. The author also desires to thank the American Council of the Institute of Pacific Relations, its secretary, Mr Frederick V. Field, and other members of the staff for making possible the publication of the book. To his wife, without whose interest and encouragement the book would never have been written, the author's indebtedness is beyond acknowledgment, and to her the book is dedicated.

CH'AO-TING CHI

BROOKLYN, NEW YORK
February 1935

CONTENTS

CHAPTER IV

ORIGIN OF WATER-CONTROL AS AN ECONOMIC FUNCTION OF THE CHINESE STATE

CHAPTER V

THE LOESS REGION AND CENTRAL HUANG HO BASIN AS KEY ECONOMIC AREA

CHAPTER VI

THE TRANSITION FROM THE HUANG HO BASIN TO THE YANGTZE VALLEY

CHAPTER VII

THE ECONOMIC DOMINATION OF THE YANGTZE VALLEY

BIBLIOGRAPHY

(With Notes on Important Chinese Sources)

ILLUSTRATIONS

KEY ECONOMIC AREAS IN CHINESE HISTORY

CHAPTER I

THE CONCEPT OF KEY ECONOMIC AREA AND ITS RELATION TO PUBLIC WORKS FOR WATER-CONTROL [1]

General Introduction

"When water-benefits are developed, there will be good results in agriculture, and when there are good results in agriculture, the state's treasury will be enriched." [2] Thus the distinguished Minister Mu Tien-yen of the Ch'ing or Manchu dynasty, in a memorial to the throne submitted in 1671, put in a nutshell the intimate connection between water-control, agricultural productivity and the state of the public treasury in China. The history of this type of public works, the records of which extend over centuries, is of unique value in estimating the economic basis of unity and division in Chinese history.

The development of "water-benefits," or the construction of water-control works for the sake of increasing agricultural productivity and facilitating transportation, especially the transportation of grain tribute, was, in China, essentially a function of the State. The fact that irrigation canals, surface tanks, drainage and flood-control works and artificial waterways were mostly built as public works links them

[1] The substance of this chapter was presented before the Far Eastern Section of the one hundred and forty-sixth meeting of the American Oriental Society, Philadelphia, 4th April 1934, and published in *Pacific Affairs*, vol. vii, No. 4, December 1934, under the title "The Economic Basis of Unity and Division in Chinese History."

[2] *Gazetteer of the Prefecture of Suchou* (Soochow, in Kiangsu Province), *chüan* 11, pp. 3-4.

closely with politics. They were used by successive dynasties as an important political lever and a powerful weapon in social and political struggles. The objectives and the development of such public works were decided primarily not by humanitarian considerations but by natural and historical conditions, and the political objectives of the ruling classes. In each period of Chinese history certain regions received more attention than others. Each region of this kind was a favoured area developed by the authorities at the expense of other regions for the purpose of maintaining or building up what may be called a *Key Economic Area*.

By using the concept of a Key Economic Area, it is possible to analyse the function of the economic base as providing the fulcrum for the political control of subordinate economic areas in China. It thus becomes possible to study an important aspect of Chinese economic history, to approach it from the viewpoint of political power with reference to regional relations, and to formulate it in terms of the development of agricultural productivity by irrigation and flood-control and the evolution of a system of artificial waterways, primarily for the transportation of grain tribute to the seat of government. No other method reveals so clearly the relation of political power to geographical regions in China, the recurrent dominance of one region over others, and the means by which political unity in a large territory of marked regional diversity was achieved. It is important to point out in this connection that regional economic geography has influenced the history of the peasant revolts of China (which were often the cause of a change of dynasty), both by providing a focus of rebellion and by restricting or augmenting the chances of success.

With the passing of classical feudalism in the third century B.C., China entered on a long period characterized by territorial expansion and a shifting of the economic focus, together with alternations of political control, but practically without change in either the character of society or its political superstructure. This period did not in fact end until the breaking down of Chinese isolation in the middle of the nineteenth century. It therefore includes practically the whole of Chinese history, from the beginning of the imperial or dynastic form of unification to the invasion of China by the Western Powers. Within this period, two outstanding movements can be observed. One is the alternation of unity and division effected on a more or less unvarying plane of socio-economic development, with an almost complete lack of structural change, which has often been interpreted as stagnation. The other is the southward movement of Chinese civilization from the central Yellow River basin to the Yangtze valley, which is distinctly a phenomenon of growth. In the course of the advance from the central Yellow River basin to the Yangtze valley, the relative strength and political importance of the various regions changed progressively, with a corresponding change in the Key Economic Area as the central problem of regional control. Thus, isolating the phenomenon from the influence of invasion, peasant revolts, the development of commerce andother factors, the question of territorial expansion, together with changes in the economic and political centre of gravity, but without structural change in either social or economic forms, becomes a problem of the shifting of Key Economic Areas, the solution of which provides an important key to the understanding of Chinese history.

Definition of Key Economic Areas

Chinese economy throughout the long period under discussion was primarily composed of tens of thousands of more or less self-sufficient villages which were normally woven into larger groups for purposes of administration or military action. The larger unit of political administration, equivalent to the modern "province," has existed since the Han dynasty (206 B.C. to A.D. 221). The name has been changed under different dynasties and the boundaries of individual provinces have been changed from time to time ; but the provincial unit, as such, has continued almost unchanged from very early times. These provincial groupings, however, were again combined into geographical regions, according to major topographical divisions, and through economic factors. The outline of geographical regions of this kind was particularly emphasized in times of disturbance and divided rule. A comparison of maps in the disturbed years between the Ch'in (221-206 B.C.) and Han (206 B.C. to A.D. 221) dynasties or between the Sui (589-618) and T'ang (618-907) dynasties, for instance, with maps in the equally disturbed periods of the Three Kingdoms (221-264), the Northern and Southern Dynasties (420-589) and the Five Dynasties (907-960) will bring this fact out very clearly.[1] Commercial growth in China never reached a level which would enable it to overcome the localism and narrow exclusiveness of an agricultural economy. These regional groupings were highly self-sustaining and independent of each other, and in the absence of machine industry, modern

[1] See Ou-yang Yin, *Historical Atlas of China*, Wuchang, China, 1930 (in Chinese).

facilities of transport and communication and an advanced economic organization, state centralization in the modern sense was impossible. In the circumstances, the unity or centralization of state power in China could only mean the control of an economic area where agricultural productivity and facilities of transport would make possible the supply of a grain tribute so predominantly superior to that of other areas that any group which controlled this area had the key to the conquest and unity of all China. It is areas of this kind which must be designated as the Key Economic Areas.

The Function of Grain Tribute

The vital importance of the grain tribute can be seen from the fact that in Chinese history taxes were largely collected in kind. The grain tribute was the chief source of supply for the imperial clan, the central bureaucracy, and the army concentrated at the seat of power. This function of the grain tribute is very clearly explained in a preface to the handbook on the Grand Canal by Tung Hsün-he, a high mandarin in the Ch'ing or Manchu dynasty. He says:

> The capital stands at the upper [northern] part of the domain. The demands for [State] worship and for the supply of the court, the salaries of officials and stipends for scholars, and, above all, the commissariat for the army, all depend upon the grain tribute. Civil and military officials, and their servants and soldiers whose names are registered in the commissariat records, number 170,000. Assuming that each has a family of eight persons, the total number of people dependent upon the grain tribute [outside of the demands for worship and for the supply of the court] would be 1,360,000. Even if the calculation is made at the rate of five persons in each family, the number would total 850,000. It certainly would not do merely to buy several hundred thousand *ho* from several tens

of hundreds of villages. . . . The demand can only be met by transporting grain tribute from the south-east.[1]

Tung was referring to the specific situation in the Ch'ing dynasty, but the general principle of the statement is applicable to practically every other dynasty since the Ch'in (221-206 B.C.).

Aside from supplying the needs of the capital, the grain tribute was also the source for the accumulation of an indispensable reserve, particularly in order to prevent possible rebellion, or to feed a large force assembled for the purpose of suppressing a rebellion, in case the preventive measures had proved to be ineffective, or of fighting a foreign war in case of invasion. As far back as the Han dynasty, Chia Yi (200-168 B.C.), the able adviser of Han Wu Ti, realized the political and military significance of such a reserve. "Both public and private reserves are in a pitiable condition, although the Han dynasty has been in existence for forty years," he remarked to his master, and proceeded to ask, admonishingly, "In case of a famine in a territory of two or three thousand *li*, where could relief be procured? And if there are emergencies on the border, how could the provisions for several tens of thousands of soldiers be supplied?"[2] A sufficient supply of provisions has always been a vital concern for any army, but before the advance of modern warfare it is no exaggeration to say that food was the life of an army and a sufficient grain reserve its most important weapon.

Both functions of the grain tribute are well

[1] Tung Hsün-he, *Chiang-pei Yung-chen*, or *Handbook on the Course of the Grand Canal North of the Yangtze*, pp. 1-2.

[2] Chia Yi, *Hsin Shu*, or *New Book*, *chüan* 73. The words quoted can also be found in Pan Ku, *Ch'ien Han Shu*, or *History of the Earlier Han*, *chüan* 24, Book on Food and Commodities, p. 9.

summarized by a Sung scholar in an introduction to a chapter on the transport of grain tribute in the famous *Ts'e-fu Yüan-kuei*, completed in 1013. It states positively that

> The capital is where a large mass of people concentrates, and where ten thousand troops and hundreds of officials look for supplies. How can it be possible for tax resources of the capital area to be sufficient to meet the demand? When it comes to putting down rebellions by imperial order, either advancing far into the enemy's territory, which necessitates the carrying of an abundance of provisions, or following up the enemy's movements, which must be supported by a light commissariat, [the army] requires transport of grain tribute to supply its needs.[1]

In the circumstances it is small wonder that the grain tribute, its production, collection and transportation, has always been the chief concern of the ruling house and its bureaucracy.

Economic Basis of Unity and Division

The task of remitting annually a part of the local revenue as grain tribute was considered one of the main duties of local officials. Definite quotas were assigned to the various provinces during the Ch'ing dynasty,[2] for instance. But in times of disturbance, only the area under the firm and direct control of the central government could be depended on for continued remittance of the tribute. Local officials or self-appointed chieftains in those areas, which the power of the central government could not easily reach, would take advantage of the disintegration of the ruling

[1] Wang Ching-jo and Yung Yi, *Ts'e-fu Yüan-kuei*, or *Anthology of Historical Records*, *chüan* 498.

[2] Huang Han-liang, *The Land Tax in China*, N.Y., Columbia Univ. Press, 1918, pp. 92-95.

dynasty and rule these territories independently. When the areas occupied by these chieftains happened to be economically of equal strength, the objective material condition for a sort of balance of power existed, and when other factors did not upset the balance, there would be a protracted period of division. During periods of division, which inevitably involved struggle, rival rulers not infrequently resorted to the construction of public works for water-control. This competition in constructive activity, going on simultaneously with the wanton destruction common to feudal wars, generally ended with an upset of the balance and the creation of a new dominant economic area. The events during the last years of the Three Kingdoms, a classical example of division, when the complicating factor of nomadic invasion was absent, offer a noteworthy example. "From Huang Ch'u (221-226) to the Tsin dynasty (265-419), able ministers all considered the cutting of canals and storing of grain the means for military preparations." [1]

Whenever such a dominant economic area was in existence, the chieftain who seized control of the Key Economic Area obtained a predominant material advantage over the other contending groups and could eventually put the country under one rule. With unity thus achieved, the ruling group, in order to maintain its power, generally paid special attention to the further development of the agricultural productivity and transport facilities of this Key Economic Area. The study of Chinese history from the standpoint of the Key Economic Area as a lever of control will therefore throw much light on the central question of unity and division. It will also provide a guide to

[1] K'ang Chi-t'ien, *Ho-ch'ü Chi-wen*, or *Notes on Rivers and Canals*, *chüan* 4.

the understanding of the basic objective conditions that determined the economic policy of the various dynasties in the whole period of Chinese history, from the beginning of the Ch'in dynasty in 221 B.C. to the end of the Ch'ing dynasty in 1912.

SHIFTING OF KEY ECONOMIC AREAS IN CHINESE HISTORY

The economic history of China from 255 B.C. to A.D. 1842 (the beginning of the modern period of foreign impact), judged by this criterion, can be divided into five periods. *The first period of unity and peace* covers the Ch'in and Han dynasties (255 B.C. to A.D. 221), with the Ching, Wei, Fen, and lower Huang Ho (Yellow River) valleys as the Key Economic Area. *The first period of division and struggle* (a most important transitional period) covers the Three Kingdoms, the Tsin dynasty and the Southern and Northern dynasties (221-589), with Szechwan and the lower Yangtze valley, gradually developed by irrigation and flood-control, emerging as important areas of agricultural production to challenge the dominance of the Key Economic Area of the earlier period. *The second period of unity and peace* covers the Sui and T'ang dynasties (589-907), with the Yangtze valley assuming the position of a Key Economic Area and the simultaneous rapid development of Grand Canal transportation connecting the capital with the new Key Economic Area. *The second period of division and struggle* covers the Five Dynasties, the Sung dynasty and the northern dynasties of the Liao and Kin (907-1280), with additional intensive development of the Yangtze valley as the outstanding Key Economic

Area in China. *The third period of unity and peace* covers the Yüan, Ming and Ch'ing dynasties (1280-1912), with the rulers increasingly worrying over the distance between the capital and the Key Economic Area, and repeatedly attempting to develop the Hai Ho valley (now Hopei province) into a Key Economic Area.

These five periods represent stages in the long-term change in Chinese socio-economic history, marked by the shifting of Key Economic Areas from one region to another. It goes without saying that in each period there were short intervals of social and political disturbances, frequently originating in peasant revolts, which usually resulted in the replacing of one dynasty by another, as in the downfall of the Yüan dynasty and the foundation of the Ming dynasty in 1368. Other interruptions were caused by barbarian invasions which were usually encouraged by internal economic breakdown in China such as those of the Khitan Tatars (Liao dynasty, 916-119), the Nüchen or Juchen Taters (Kin dynasty, 1115-1260), the Mongols (Yüan dynasty, 1280-1368) and the Manchus (1644-1912). These short cycles, however, can be better understood if they are regarded as subordinate to the long-term cycles governed by the shifting of the economic centre of gravity, which provided the setting for political and dynastic movements, whether they took the form of internal rebellion or of alien invasion.

Thus the general line of development of public works for water-control can be traced to, and understood in terms of, the necessity of holding and developing the Key Economic Area as the principle underlying the economic policy of succeeding dynasties. By clarifying the course of development of public works for water-control, the concept of the Key Economic Area illuminates a most significant feature in the historical processes of the whole semi-feudal period of Chinese history.[1]

[1] Dr K. A. Wittfogel speaks of "the economic-political kernel-district of China" in his monumental work on Chinese economy and society. His contribution, however, is different from mine, although our theses are complementary to each other. He says, "The so-called 'kultural centre,' or more correctly, the economic-political kernel-district of China, was by no means situated always at the same spot. It shifted several times during the period when China was predominantly agricultural; an industrialized China would create new centres of power at new spots, because the centres of raw materials and of production of industry mostly do not coincide with the agricultural centres of production" (*Wirtschaft und Gesellschaft Chinas*, p. 273). On the basis of the shifting of the kernel-district, which he ascertained to have occurred three times, from the north-west to the north-east and then the Yangtze Valley, he formulated the conception of three stages of the development of Chinese culture. Thus his contribution lies in having brought out the fact of the changing geographical location of what he calls the kernel-districts and their relation to Chinese culture. The central idea of my work, on the other hand, is to explain the function of the Key Economic Area as an instrument for the control of subordinate areas, to indicate the manner by which the shifting of the Key Economic Area was brought about, and to provide an explanation of the economic basis underlying the alternate occurrence of unity and division in Chinese history. I am indebted to my friend Joseph Pachtman for orally translating to me part of Dr Wittfogel's *Wirtschaft und Gesellschaft Chinas*, and to Dr Wittfogel for providing me with the English text of the above quotation and other excerpts from his book.

CHAPTER II

GEOGRAPHICAL BASIS OF WATER-CONTROL AND THE ECONOMIC REGIONALIZATION IN CHINA

GEOGRAPHICAL conditions in China are such that without the development of a system of water-control as an integral part of its agricultural practice, agricultural production could never have reached so high a stage as it did. The flourishing culture of semi-feudal China, which was the outgrowth of a highly productive agricultural economy, would have been impossible. The vital importance of irrigation to Chinese economy has been clearly brought out by the detailed investigations of K. A. Wittfogel which convincingly establish the fact that irrigation is " everywhere in China an indispensable condition for intensive agriculture, on the basis of which Chinese agrarian society had been constructed, just as the industrial society of modern capitalism has been constructed on the basis of coal and iron." [1]

Thus, although China has many different geographic regions and each has its own peculiar characteristics, almost all the major regions demand water-control in one form or another as a basis of agricultural development. In the loess region in the north-west, the problem is primarily one of canal irrigation ; in the Yangtze and Pearl River valleys, it involves the continuous draining of the fertile but swampy alluvial land and the maintenance of a complicated system of drainage and irrigation ; while in the lower Yellow and Hwai River valleys the problem is essentially one of flood-control. In the sphere of transportation,

[1] K. A. Wittfogel, *Wirtschaft und Gessellschaft Chinas ; versuch der wissenschaftlichen analyse einer grossen asiatischen agrargessellschaft.* Leipzig, 1931, p. 229.

water routes have always been of vital importance as arteries of commerce and administration throughout the Chinese domain.

THE LOESS AND IRRIGATION

To start with "the most important cradle of early Chinese civilization,"[1] which, according to Dr V. K. Ting, is the area enclosed between lat. 31°-40° N. and long. 113°-118° E., embracing the provinces of Shansi, Honan, and part of southern Hopei, western Shantung, and northern Kiangsu and Anhwei, it is important to note that it contains more loess than alluvium. The whole Hwai valley, west of long. 118° E., is in the loess region. The area enclosed between lat. 32°-34° N. and long. 114°-118° E. is almost entirely of loess formation, except for the channel of the old Yellow River. Within the same longitudes the land between 35°-36° N. latitude is again largely of loess, except for the river valleys. Thus the peninsula of Shantung is joined to the western loess region by two belts of the same formation. The alluvium in the whole region is confined to the territory north of lat. 36° and to the river valleys.

"It is this continuous semi-steppe stretching from the sea to Turkestan free from both forest and marsh, and favourable to agriculture and to wheeled vehicle," V. K. Ting concludes, "that made early settlement and continuous diffusion of culture possible."[2] The loess area as described here is so extensive that even if we take the Chou period (*c.* 1050-255 B.C.), a later period than that taken by Dr Ting, which is probably

[1] V. K. Ting, "Prof. Granet's 'La Civilization Chinoise,'" *The Chinese Social and Political Science Review*, vol. xv, No. 2, Peiping, July 1931, p. 268.
[2] *Ibid.*, pp. 268-269.

the Yin period (*c.* 1401-1050 B.C.) and include Shensi within the cradle of Chinese civilization, the whole cradle would still lie in the loess territory.

This description of the outstanding geographical feature of the territory of ancient China provides a key to the understanding of the vital importance of irrigation to Chinese agriculture. The secret, of course, lies in the special character of loess in relation to water supply.

Professor E. F. von Richthofen's investigations revealed this fact about half a century ago.[1] He observed that the loess receives water like a sponge. The high porosity and great capillary capacity of the loess which enables the mineral elements hidden in deep soil to rise to the top, thus bringing them within reach of the roots of the crops, endow the loess with the capacity of self-fertilization. However, it is evident that this mechanism can only be brought into action when there is sufficient water. This is why Lyon, Fippin and Buckman conclude, in their authoritative book on *Soils: Their Property and Management*, that "Whenever moisture relations are favourable loess is an exceedingly fertile soil, due to its rich stores of potash, phosphorus, and lime."[2]

From this, George B. Barbour, formerly Professor of Geology at Yenching University, Peiping, China, concludes that "since the Chinese loess often already carries a high lime-percentage and the practice of manuring is universal, water supply becomes the vital factor."[3] These are not merely *a priori* conclusions.

[1] E. F. von Richthofen, *China*, Berlin, 1882, vol. ii.

[2] T. Lyttleton Lyon, Elmer O. Fippin, and Harry O. Buckman, *Soils: Their Property and Management*. New York, 1915, p. 61.

[3] George B. Barbour, "The Loess of China," *The China Journal of Science and Arts*, vol. iii, No. 8, Shanghai, August 1925, p. 462.

They are supported by the capillary experiments made by Professor T. New of Tsing Hua University, Peiping, China,[1] the water-holding capacity of loess as determined by Professor Lowdermilk of the University of Nanking, Nanking, China,[2] and the chemical analysis of loess worked out by Dr W. H. Wong of the Geological Survey of China.[3] Professor Barbour reinforced his conclusions by the remark, undoubtedly based upon personal observation, that " these theoretical conclusions seem to be borne out by the facts wherever observed in the field." [4]

Value of Silt as Fertilizer

As for the agricultural fields which are situated in alluvial plains, river valleys, or former river or lake beds, where the best agriculture in North as well as South China is found, irrigation as a fertilizing agent is just as important, if not more so, as in the loess steppes, though it operates differently. Here the result is largely achieved by flooding. Even to-day, it is not an uncommon thing to witness, in North China, great torrents of silt-laden water rushing down the hills after a summer storm. Canals and ditches are constructed by the peasants, and, if on a larger scale, by the government, in order to capture the mud-laden water and guide it to the fields, for the treble purpose of irrigation, fertilization and flood-prevention.

The big rivers in North China, such as the Ching and Wei (in Shensi), the Fen (in Shansi), the Lo (in Honan) and the Yellow River, all collect such torrents as they flow along their course, and, as a result,

[1] George B. Barbour, "The Loess of China," *The China Journal of Science and Arts*, vol. iii, No. 8, Shanghai, August 1925, pp. 517-519.
[2] *Ibid.*, p. 462. [3] *Ibid.*, p. 519. [4] *Ibid.*, p. 462.

practically all these rivers carry a tremendous amount of silt, which can be diverted advantageously to the fields by a system of canalization. The silt content of the Yellow River averages 11 per cent.[1] Traditional Chinese writers, following the estimate of Chang Jung of the Earlier Han dynasty (206 B.C. to A.D. 25),[2] put the ratio between water and silt carried by the Yellow River as ten to six. The Ming official, P'an Chi-hsün (1521-1595), famous for his water-control achievements and author of a classical work on water-control, stated that during the autumn, the silt content of the Yellow River increased from 60 per cent. to 80 per cent.[3] These statements obviously can only be regarded as literary exaggerations for the purpose of emphasis and can by no means be considered scientific estimates. However, they do indicate that the heavy silt content of the Yellow River has long ago attracted the attention of Chinese administrators.

There are authentic records which show that, as early as the Earlier Han dynasty (206 B.C. to A.D. 25), Chinese peasants knew the value of silt as a fertilizer. In about 95 B.C., after the completion of the Po Canal, which conducts the silt-laden water of the Ching River to a large agricultural area in the heart of modern Shensi, the peasants of the region sang in its praise. The song contains the following lines :[4]

> A *tan* of Ching water contains much silt ;
> It irrigates and it fertilizes ;
> It makes your crop grow ;
> It feeds millions in the country's capital !

[1] Chang Gee-Yuen, *Geography of China*, Chungshan Textbook series, vol. i, Nanking, 1932, p. 7.
[2] Pan Ku, *Earlier Han History*, *chüan* 29, Book on Canals and Ditches, p. 16.
[3] *The Gazetteer of Honan Province*, 1869, *chüan* 14, p. 23.
[4] Pan Ku, op. cit., *chüan* 29, Book on Canals and Ditches, p. 8.

Chang-sun Wu-chi, the famous T'ang minister under T'ai Tsung (627-649) and Kao Tsung (650-685), once said of the same canal, " the water of the Po canal contains silt. It irrigates the field and adds to its fertility."[1] During the Ming dynasty, P'an Chi-hsün (1521-1595) noted that one of the reasons for the breaking of the dike of the Yellow River was that certain peasants secretly cut the dike in order to enrich their exhausted fields.[2]

In speaking of the Yin River in Honan Province the local gazetteer of Linyin District, published in 1660, contains a very illuminating paragraph. It reads :

> The source of the Yin river is deep and small, but after heavy rain during the summer and autumn seasons, it usually swells in volume, breaks its dikes, and is very difficult to control. But wherever the mud-laden water passes, the area becomes a fertile country. Because of the terrain people who live in the south-west are benefited by it, but those who live in the east and north-east suffer damages. If, for several years, the river fails to break its dikes, the fields become thin ; the peasants, then, will secretly cut the dikes when the river rises and thus enrich their land. The people of Yin (Linyin district in Honan province) thus speak of the Yin River as both harmful and beneficial.[3]

To take a very recent case, the flood caused by the overflowing of the Lo River in Honan Province in July 1932 was the worst in eighty years. It destroyed many villages ; but the wheat crop in 1933 was unusually good.[4] In another case many hundred *mu* of land in T'ungchou in Shensi Province, after being flooded in 1932, produced a crop that had not been seen in that locality for many years.[5]

[1] *Gazetteer of Shensi*, 1735, *chüan* 39, p. 64.
[2] *P'an Chi-hsün*, Handbook for River-control, 1590, vol. iv, p. 45.
[3] *The Gazetteer of Linyin District, Honan Province*, 1660, vol. i, p. 7.
[4] *Ta-Kung-Pao* (a leading daily in North China), Tientsin, 3rd Mar. 1933.
[5] *Ibid.*

Mr Li She (Hsieh), a leading water-control expert and former chairman of the National Yellow River Conservation Board, took clear recognition of this fact and proposed that a system of irrigation canals and ditches should be constructed, and that when the silt deposited at the bottom of the canals and ditches is dug out during the process of deepening the canal at regular intervals, it be used to enrich the fields.[1]

There are two interesting cases recorded in the *Gazetteer of Shansi*, published in 1887, which bear witness to the fertility value of silt in that province. In one case the Gazetteer speaks of irrigation by canals and ditches fed by the Yu River in Tat'ung district, and states that "after one or two years of silting through irrigation by the waters of the Yü River, sandy and gravel wastes were turned into fertile land."[2]

In a memorial to the throne submitted in 1729, the mandarin Han Kuan-ch'i referred to the case of the Su River which flows from Chiang district, passing through Wenhsi, Hsia, Anyi, and Yishih districts, into the Yellow River. All along its course, through the five districts, peasants utilize its waters for irrigation. "The water of the Su River is deep and thickly laden with silt," the memorial reports, "and when it is used for irrigation just before and after every freezing season, the fields become doubly enriched. Therefore ignorant peasants often secretly cut the embankments and construct dams across the river [thereby inflicting damage on the peasants living down-stream]. . . ."[3]

[1] *Ta-Kung-Pao* (a leading daily in North China), Tientsin, 2nd Oct. 1933.
[2] *The Gazetteer of Shansi Province*, 1887, *chüan* 68, p. 26.
[3] *Ibid.*, *chüan* 67, p. 30.

These cases illustrate most clearly the double function of irrigation.

In Hopei province, the fact that the destruction caused by the frequently flooded Yungting River is usually compensated by the fertilizing effect of its silt deposit is generally recognized. Ch'en Yi (1670-1742) of the Ch'ing dynasty, in a famous discussion on the question of river-control, speaks of "the richness of silt deposits" of the Yungting River and definitely states that "the loss of the autumn crop because of flood was doubly made up by an unusually good summer crop."[1] In a discussion on the Yungting River, Ch'en Hung-mou (1696-1771) also refers to the exceptionally good wheat harvest after every inundation of the fields by the water of the Yungting River.[2]

The 1871 edition of the *Gazetteer of Ch'i-p'u* quotes from an earlier edition of the Gazetteer a very significant remark about the Yungting River. It says, "The thick mud of the Yungting River fertilizes crops, and whenever it accumulates, thin land is fattened and the yield increased many times."[3] In a letter discussing the problem of the control of the Hun or Sangkan River, Fang Pao (1668-1749) and Ku Yung-fang revealed that when the river passed Kuan and Pachou its embankments disappeared and spread a sheet of water one or two hundred *li* wide over these two districts. When the water receded, in ten days or so, the silt-covered fields were called by the peasant inhabitants "a land covered with gold."[4] The Chang River which is famous for its ancient irrigation works, is also known for its fertilizing faculty, as testified by Sun Chia-kan (1682-1752), one of the most distinguished officials of

[1] *The Gazetteer of Ch'i-p'u* (Hopei), 1910, *chüan* 82, p. 15.
[2] *Ibid.*, *chüan* 82, p. 56. Also see p. 63.
[3] *Ibid.*, *chüan* 85, p. 14. [4] *Ibid.*, *chüan* 82, p. 54.

the Ch'ing dynasty, in a memorial to the throne, which is also included in this Gazetteer.[1]

In South China, water-control also involves the twofold problem of regulating water supply and augmenting soil fertility, although the problem there presents itself in a form very different from that in North China. The question is not one of inundation canals carrying silt to enrich the fields, but one of the drainage of surplus water and the utilization of drained swamps and lake bottoms for the cultivation, mostly, of rice.

The practice, among powerful landowners and rich farmers of South China, of extending their fields by encroaching upon the land covered by lakes and rivers, is undoubtedly partly due to the struggle for land which, nominally belonging to the state, has not yet been claimed by any individual, and partly due to the desire to avoid taxation, which takes time to reach land newly recovered from lake bottoms or river beds.

This tendency, which gives rise to a most difficult economic and political problem which has been of special importance since the Southern Sung dynasty (1127-1280), is much stimulated by the great fertility of newly reclaimed land. It is not merely a question of struggle for public and tax-free land, but also of a struggle for fertile land. In an essay on the Pinlü lake in Ch'anglo district in Fukien province, published about 1829, the Ming magistrate Chiang Yi-chung refers to the fertility of lake land in very concrete terms. He says, "The lake bottom is largely covered with fertile soil and can be cultivated. The yearly yield on such soil is treble that of ordinary fields. Yet the government does not tax such land. The powerful and the wicked covet it and exploit it as soon as they can

[1] *The Gazetteer of Ch'i-p'u* (Hopei), 1910, *chüan* 82, p. 31.

take possession."[1] These rich "lake-and-river lands," usually reclaimed by being surrounded with dikes resembling walls to keep the water out, are called *Wei*-land, or *Yü*-land, in the lower Yangtze valley (Kiangsu, Chekiang, Kiangsi, and Anhwei), and *Yüan* in the great lake region, especially Hunan. They make up a very important proportion of the most productive land in the richest agricultural section in China.

Many outstanding Western scholars have studied and recognized the place of silt in Chinese agriculture. The amount of silt brought down by torrents from the hills or high levels in Shansi province has been measured by Professor W. C. Lowdermilk. He took six flood-water samples of 1000 cubic centimetres each from three different inundations after the crest of the flood had passed, weighed the silt after it was filtered out and dried at 100° C., and found that the amount of silt, by weight, varied from 14 to 22 per cent. The comment of the scientist deserves special attention. He says, "This is a startling amount; it indicates that flood-waters are carrying down from the mountains every season thousands of tons of productive soil layer."[2]

It is undoubtedly true that this process of soil erosion is responsible for the rocky and sterile mountains and barren wastes in the mountainous areas of Shansi and other parts of China. The importance of checking erosion as a step toward the conservation of resources for purposes of irrigation is very ably brought out in the following quotation from an

[1] *The Gazetteer of Fukien Province*, 1829, *chüan* 33, p. 13.
[2] W. C. Lowdermilk and J. Russell Smith, "Notes on the Problem of Field Erosion," *Geographical Review*, vol. xvii, No. 2, New York, April 1927, p. 227.

essay by the Ming scholar, Yen Sheng-fang, published in 1887:

Before the reign of Cheng Te (1506-1521) [during the Ming dynasty], flourishing woods covered the south-eastern slope of the Shangchih and Hsiachih mountains [in Ch'i district in Shansi province], which were not stripped because people gathered little fuel. Springs flowed into the Panto stream and flowed in long waves and powerful sweeps through the villages of Luchi and Fencha and entered the Fen River at Shangtuanto as the Changyüan River. . . . It was never seen dry any time of the year. Hence villages from afar and villages in the north of the district all cut branch canals and ditches which irrigated several thousand *ch'ing* of land. Ch'i thus became prosperous. At the beginning of the reign of Chiaching (1522-1566) people vied with each other to build houses and wood from the southern mountains was cut without a year's rest. The natives took advantage of the barren mountain surface and converted it into farms. Small bushes and seedlings in every *chih* (foot) of ground were uprooted. If heaven sends down a torrent, there is nothing to obstruct the flow of the water. In the morning it falls on the southern mountains; in the evening, when it reaches the plains, its angry waves swell in volume and break embankments, causing frequent changes in the course of the river. . . . Hence Ch'i district was deprived of seven-tenths of its wealth.[1]

But, while emphasizing the necessity of checking soil erosion, it is important to recognize the fact that the silt carried down by erosion need not be lost and may be turned to good account by a well-constructed and efficiently maintained system of canalization and irrigation. In this connection the observations of Mr Walter Mallory, who has extensive practical flood-control experience in North China, are particularly worth noting. He says:

The losses from a flood are recompensed, to a degree, by the enrichment of the soil resulting from a deposition of new earth from the flood-waters. Irrigation effects this improvement without the accompanying losses and its benefits are therefore two-fold.[2]

[1] *Gazetteer of Shansi Province*, 1887, *chüan* 66, p. 31.
[2] W. H. Mallory, *China: Land of Famine*, New York, 1926, p. 148.

Dr Louis A. Wolfanger, of Columbia University, recognizes a fact of fundamental importance, especially in regard to the question of so-called " permanent agriculture " in China, in the following comment on soil condition in China :

> [The Chinese] people made little or no use of their mature upland soils, . . . but are crowded along the rivers. The floods are tragic ; but the soils are young, fresh, and tillable and are periodically renewed. The natural productiveness is further increased by the extensive use of fertilizers. The population outside of flood plains is in hill and low mountain country, whose slopes have been laboriously carved into terraces. Their soils are also periodically immature. They were constantly rejuvenated in their natural state by erosion, and if man's use has tended to decrease their fertility it has probably been offset to some degree by the general use of fertilizers. . . . The Chinese had, and apparently still have, young, productive and unleached soils.[1]

The Physical Basis of " Permanent Agriculture "

Professor Vladimir G. Simkhovitch significantly remarks in his article, " The Fall of Rome Reconsidered," that " the experience of China and Japan has proven that on very small land-plots such intensive agriculture can maintain itself indefinitely without any recourse to scientific repletion of the soil by mineral fertilizers." He then asks an important question, the answer to which solves one of the fundamental problems in the economic history of China, the problem of " permanent agriculture " : " Why did Rome fail where China and Japan succeeded, after a fashion ? "[2]

[1] Louis A. Wolfanger, " Major World Soil Groups and Some of the Geographic Implications," *Geographical Review*, vol. xix, No. 1, New York, January 1929, pp. 106-107.

[2] V. G. Simkhovitch, *Toward the Understanding of Jesus and Other Historical Studies*, New York, 1921, p. 111. Note Essay on " Rome's Fall Reconsidered."

The answer, as the foregoing discussion indicates, lies in the self-fertilizing capacity of the loess when there is sufficient water, and the capacity for self-renewal of the rich silt deposits in the alluvial plains, which are constantly rejuvenated by erosion, either through controlled irrigation or natural flooding. In the latter case the benefit is counterbalanced by damage, but, in the long run, it keeps the soil young and productive and free from the danger of a relatively permanent depletion. This is the reason why China could stand many centuries of intense agriculture without recourse to scientific repletion of the soil by mineral fertilizers.

However, it would be a serious mistake to assume that this natural advantage of the soil condition is sufficient to guarantee a more or less even level of agricultural productivity in China. In other words, the maintenance of these favourable soil conditions, which free China from the danger of relatively permanent depletion of the soil, does not free her from the menace of a drastic decline in agricultural productivity, especially on a regional scale, as the result of certain socio-economic and political conditions.

The very advantages of the natural fertility of the loess and alluvial deposits cannot be fully realized without an efficient system of irrigation, which is an engineering as well as a socio-economic problem. The practical importance of a given region to the ruling power might lead to its rapid development and careful maintenance, but once a region fell from favour it would be neglected and abandoned to the fate of soil depletion and drastic decline in productivity as in the case of Shensi province during the Sung (960-1280), Yüan (1280-1368), Ming (1368-1644), and Ch'ing (1644-1912) dynasties. Favourable natural conditions provide a basis for water-control development, which conditions and is conditioned by the location of the Key Economic Area and the general socio-economic and political conditions of the time.

VARIABILITY OF RAINFALL IN CHINA

The special character of the soil in China discussed above makes it evident that water supply is a vital factor in agricultural development. The next question is, How far can the natural process of annual rainfall be depended upon for supplying this vital necessity? Although the severity, duration and continuous recurrence of drought and flood throughout Chinese history have been much aggravated by socio-economic causes, nevertheless these catastrophes do indicate that the natural process of water supply in China is not only unreliable but positively disastrous.

Commonsense observation, as well as scientific opinion, seems to point to the conclusion that the main trouble with the rainfall in China, especially in North China, does not lie in its general abundance or deficiency, but in its great variability from year to year. The rainfall of China, like that of India, is brought by the monsoon winds which change with the seasons.[1] The prevailing direction of the winds is sharply reversed from summer to winter. The following statistics show the degree of completeness of the wind reversal in North China:

North China Percentage Wind Frequency[2]

	N.	N.E.	E.	S.E.	S.	S.W.	W.	N.W.
Winter . . .	17	8	5	6	6	8	18	32
Summer . . .	10	9	12	26	16	10	7	10

[1] Julius Harm, *Handbook of Climatology*, Part I, "General Climatology" (translated by Robert de Courcey Ward), New York, 1903, p. 163.

[2] W. G. Kendrew, *Climate*, Oxford, 1930, p. 97.

A glance at this table is sufficient to impress the observer with the importance of the climatic rhythm in Chinese agriculture. It is equally clear, as Kendrew states, that "any serious departure from the normal weather may lead to great distress and even widespread famine."[1] But, "unfortunately," as Kendrew further states, "there are large variations in both the duration and the amount of the rain, especially in India and China."[2] The most recent authoritative works on the subject show complete agreement as to the large degree of the variation of the rainfall from year to year in China. In his comprehensive study on the monsoon, the French meteorologist, Jules Sion, emphatically points out the great fluctuations of rainfall from year to year as one of the characteristics of the monsoon climate.[3] A. Austin Miller also states that "monsoon rainfall, largely cyclonic in origin, is characteristically variable in amount, and famine is the curse of monsoon lands."[4]

But natural deficiencies usually stimulate human effort and thus set counteracting influences in motion. Hence, unfortunate as this characteristic of the monsoon is, "it supplied," as Miller correctly indicates, "a valuable incentive towards irrigation, in its turn a powerful stimulus to the growth of a Chinese civilization."[5]

[1] W. G. Kendrew, *Climate*, Oxford, 1930, p. 150.
[2] *Ibid.*, p. 150.
[3] Jules Sion, *Asie des moussons*, Paris, 1928-29, 2 vols. See the review by Frederick Hung in *The Chinese Social and Political Science Review*, vol. xvii, No. 2, Peiping, July 1933, pp. 360-366.
[4] A. Austin Miller, *Climatology*, London, 1931, p. 200.
[5] *Ibid.*, p. 200. Miller's economic deductions on this subject, however, are not all justifiable. He says, on the same page, "the very fertility of the lands encourages the growth of population up to the limit of supporting power, and agricultural practice is all too frequently based on the most optimistic expectation of rain. For this reason the disaster of a deficiency is all the greater, particularly in North China where the annual amount

Rice and Irrigation

Another factor which renders irrigation a vital concern in Chinese agriculture is the prevalent use of rice as the leading crop. It is a matter of common knowledge that rice cannot be cultivated without a sufficient supply, and a careful regulation, of water. An old Chinese agricultural treatise, published in the Yüan dynasty (1280-1368), describes the process very clearly. It says, "cultivators of rice build surface tanks and reservoirs to store the water, and dikes and sluices to stop its flow [when necessary]. . . . The land is divided into small patches, and after ploughing and harrowing, water is let into the fields and seeds thrown in. When the plants grow five to six *t'sun* (inches) tall, they are transplanted. All farmers now use this method. When the plants attain a height of seven or eight *t'sun*, the field is hoed, and after hoeing the water is pumped out of the fields. Then when the plant begins to flower and seed, water is again let into the fields."[1] Thus

is lowest and the margin of sufficiency least." The fallacy involved in this reasoning is that mere fertility of the lands, in the absence of certain socio-economic factors, does not encourage the growth of population. The socio-economic system of China has been such that the bureaucratic and landowning class, knowing that labour is the most important productive factor in Chinese agriculture under the prevailing level of technique, has made the encouragement of population growth one of the corner-stones of its economic policy. The "religion" of ancestor worship has its social root in this set-up. The only checks to population growth have been the limited power of sustenance of the resources remaining after the tribute to the ruling classes has been deducted and social upheavals, which usually mean a drastic reduction in the population. The surplus which the bureaucrats and landowners appropriate for their own use and extravagance has to be deducted before the average *per capita* wealth of the toiling population can have any meaning. The *per capita* wealth of the peasants, computed in this manner, would be very small indeed. Hence, in their struggle to better conditions that border on starvation, they have had to adopt agricultural practices which are "all too frequently based on the most optimistic expectation of rain." The struggle is a very difficult one. With meagre reserves, or none at all, any serious deviation in natural conditions produces a disastrous famine.

[1] Wang Cheng, *Book of Agriculture*, 1314, *chüan* 7, p. 5.

the regulation of water constitutes a most important factor in rice farming. It is safe to say that without irrigation there would be no cultivated rice.

"The Land of Rivers"

Happily, China possesses a natural advantage which greatly facilitates the work of irrigation. L. Richard, the veteran geographer of China, states that "no country in the world is so well watered as China."[1] K. A. Wittfogel brought out very emphatically the great value of the river system of China by comparing it with that of Egypt and Mesopotamia. He shows that the rivers of China do not flow through oases but are advantageously distributed over an immensely large and continuous land. In China, where peculiarities of soil and climate make water supply a vital necessity, the river system is a particularly important fertility factor. This is the reason why the centres of Chinese agriculture have up till now been situated in the big river valleys.[2]

James Fairgrieve calls China "The Land of Rivers,"[3] and elaborates his point by saying that "China is specially a land of rivers, not only in the sense that rivers flow through it, but in the sense that its history has been greatly affected by other controlling facts."[4] From the point of view of irrigation and water transportation, Fairgrieve's seemingly sweeping statement is not an exaggeration.

[1] L. Richard, *Comprehensive Geography of the Chinese Empire and Dependencies* (translated by F. M. Kennelly), Shanghai, 1908, p. 15.
[2] K. A. Wittfogel, *Wirtschaft und Gesellschaft Chinas; Erster Teil; Produiktivkrafte, Produktions-und Zerkulationsprozess*, Leipzig, 1931, p. 88.
[3] James Fairgrieve, *Geography and World Power*, London, 1917, p. 225.
[4] *Ibid.*, p. 234.

CLIMATIC PULSATION AND IRRIGATION

The factor of long-term periodical changes in climate, especially in rainfall, which have been so much emphasized in the work of Professor Ellsworth Huntington,[1] and carefully studied by Co-Ching Chu, in the case of China,[2] does not alter the conclusions of the present study. It at best merely introduces a complicating factor. It would seem probable that during long periods of aridity there should be a stronger stimulus to irrigation activity, while in times of abundant rainfall such activities should tend to decline in vigour. But the statistical data as analyzed in the following chapter [3] do not bring this out. On the other hand, the statistical data do suggest a correlation between irrigation activity and the political and economic importance of an area. This the following chapters of this study attempt to show. For the purpose of the present study it seems sufficient to point out that long-term climatic pulsations have affected the course of irrigation development in China only indirectly, in so far as they have influenced migrations and nomadic invasions.

As for the almost progressive economic decline of the cradle of Chinese civilization in the north-west (especially Shensi) during the past ten centuries, the cause clearly cannot be ascribed to long-term climatic change. Convincing evidence of a progressive dessication in this region during the corresponding period of economic decline is lacking, and a more tenable

[1] E. Huntington and S. S. Visher, *Climatic Changes*, chap. v, "The Climate of History," pp. 64-97.

[2] Chu Co-Ching, "Climatic Pulsations during Historic Time in China," *Geographical Review*, vol. xvi, New York, 1926, p. 274.

[3] Cf. *infra*, p. 35 ff.

explanation of the decline seems to lie in the development of a more fertile region in the Yangtze valley to serve as the economic base of the ruling group, and the consequent neglect of irrigation works in the comparatively unproductive province of Shensi. Even if progressive dessication be a fact, it would not refute the above contention, because it would merely aggravate the danger of the neglect of irrigation in the affected region.

Geographical Basis of Regionalization

As the natural conditions that made water-control a necessity in Chinese agriculture are the character of its soil and climate and the peculiar demands of its major food crop, rice, the geographical basis for regionalization is the peculiar character of its topography. In China, as everywhere else, the arrangement of mountains and rivers decides the major topographical divisions of the country. However, most Chinese rivers run in an east-west direction, in contrast to the north-south direction of, for instance, the rivers in the United States of North America.

The mountain ranges separating the three major river systems of China constituted barriers that created an economic and political regionalization and provided the physical basis for divided rule in China for many centuries.

As Professor Cressey puts it, "greatest of all the mountains of China is the eastward extension of the Kun Lun, known in China collectively as the Tsingling Shan, which reach eastward from Tibet nearly to the Pacific. These mountains divide China into two major geographic regions, characterized by striking contrasts in climate, agriculture, and human activities."[1] Taking into consideration merely the territory of China proper, undoubtedly the difference between the North and South, or the Huang Ho (Yellow River) valley and the Yangtze valley, constitutes the major regional difference in China, and the expansion of Chinese civilization from one of these two regions to the other brought about a decisive transition in Chinese history and marked a very crucial step in the advance of Chinese culture over the Asiatic continent. Besides this major division, there are two other important regions in China proper and many others of lesser significance that deserve consideration.

The two most sharply marked regions, aside from the Huang Ho and lower Yangtze valleys, are Szechwan-Yünnan and Kwangtung-Kwangsi. The Red Basin of Szechwan is encircled by high barrier ranges, and Professor Cressey correctly thinks that it " makes isolation a distinctive feature of the human geography." [2] Professor Cressey also reports that " the climate is favourable, the soil productive, the people are energetic and the natural resources abundant . . . and it is said that everything which can be grown anywhere in the country may be produced here." [3] Thus Szechwan, with its easily defended boundaries and rich and varied resources, is remarkably fitted for an independent and self-sufficient existence.

This is the reason why, as the versatile Chinese scholar Liang Ch'i-ch'ao observes, " whenever there were disturbances under heaven (meaning in China) Szechwan was held by an independent ruler, and it

[1] George Babcock Cressey, *China's Geographic Foundations*, New York and London, 1934, p. 38.

[2] *Ibid.*, p. 312.

[3] *Ibid.*, p. 310.

was always the last to lose its independence."[1] Such instances have occurred seven times in Chinese history since the fall of the Earlier Han (206 B.C. to A.D. 25). First it was Kung-sun Shu who became king of Szechwan, ruled that region from A.D. 25-86 and was subdued by the generals of the first Emperor of the Later Han long after the latter had unified all China. Next came the famous Liu Pei, who founded one of the Three Kingdoms, Shu (221-263), which figured so prominently in the history and literature of China. The third was Li Hsiung, who assumed the title of King of Ch'engtu at the beginning of the third century, when the Empire of the Western Tsin was dismembered.

The fourth period was that of Wang Chien and Meng Chih-hsiang, whose two houses ruled Szechwan successively as Chien-Shu (907-925) and Ho-Shu (934-965) during the period of the Five Dynasties (907-960). The fifth was Ming Yü-chen, the founder of the Ta-Hsia dynasty (1362-1371) in Szechwan at the beginning of the Ming dynasty. Then came the notorious Chang Hsien-chung, who ruled Szechwan at the end of the Ming, and the heroic Shih Ta-k'ai, who established himself in Szechwan when the Taiping Empire (1851-1863)[2] collapsed in China. Even in our day, Szechwan still seems to be a world by itself, though steam navigation on the Yangtze has penetrated irreparably its ancient wall of isolation.

As for Yünnan, Liang Ch'i-ch'ao correctly considers it to be a supplementary region to Szechwan. Hence

[1] Liang Ch'i-ch'ao, "Essay on Chinese Geography," *Yin-Pin-Shih Collected Essays*, Shanghai, 1926, *chüan* 37, p. 51.

[2] The Taiping Empire, though not recognized officially as an empire, was formed by processes closely parallel to those which formed other empires in Chinese history, and the Tai'ping Tien-kuo, or Heavenly Nation of Great Peace, as it was called, lasted longer and controlled more territory than some of the minor dynasties that are officially recognized.

Chu-ko Liang of the period of the Three Kingdoms saw the necessity of first pacifying the Yünnanese in the south before he proceeded with his plan of conquest in the north. "Politically, Szechwan and Yünnan are really one independent area."[1]

In the case of Kwangtung and Kwangsi, which were once united in the Viceroyalty of Liang-Kwang or the "two kwangs," their regional integrity can be seen very clearly from the topography. "The hydrographic drainage [of the region] centring in a single delta is paralleled by trading activities, most of which follow the waterways. Encircled by mountains and by the ocean, the cultural life of the region is self-sustained with comparatively few contacts with adjoining provinces. Climate, soils, vegetation, and agricultural practices are essentially of a kind and present contrasts to those elsewhere. The coastal fringe west of the mouth of the Si Kiang, together with the island of Hainan, are maritime and quite tropical and have less in common with the rest of Liang-Kwang, but it hardly seems proper to place them in a separate region."[2] Historically, though one of the latest regions to be developed by the Chinese, it has attained a remarkably high level of culture during the last century. The political consequences of its regional unity can find ample illustrations from recent history, particularly that of the last twenty years.

Other regions which are more or less integral units, when considered geographically and historically, are Shansi and the south-eastern coastal provinces of Chekiang and Fukien. Of the two regions, Shansi is strong in defence though weak in economic (agricultural) self-sufficiency, while Chekiang and Fukien are

[1] *Ibid.*, p. 41. [2] George Babcock Cressey, op. cit., p. 351.

weaker in defence but strong in economic resources. Historically, both have been seats of independent rulers for considerable lengths of time during periods of division. But these two regions, as well as the rest of China proper, were too close to the central domain to be able to defy the central authority for long, and, despite short intervals of independence, they were quickly swept into the whirlpool of major contentions in the Huang Ho and Yangtze River valleys. Throughout Chinese history most of the decisive battles have been fought in the Hwai River valley, the intervening strip of territory bridging the two major regional divisions in China proper.

CHAPTER III

A STATISTICAL STUDY OF THE HISTORICAL DEVELOPMENT AND GEOGRAPHICAL DISTRIBUTION OF WATER-CONTROL ACTIVITIES[1]

THE fact that the course of the development of public works for water-control was to a large extent determined by the political objective of the ruling group, which was to strengthen their hold on the country, and that this objective was attained economically by emphasizing the development of public works for water-control in a particular region, a Key Economic Area, which would serve as an economic base for the subjugation and domination of the subordinate areas, is brought out very convincingly in the accompanying "Statistical Table showing the Historical Development and Geographical Distribution of Water-Control Activities in China."

The data used are all that can be found in the provincial gazetteers and, barring accidental mistakes, there is no room for the selection of data to suit *a priori* conclusions. That the data are interpreted from the viewpoint of the subject under study and in the light of the author's particular approach must be admitted; but that is the only way statistical material can have any meaning, and when the particular approach is thus pointed out, the danger of abusing statistical data and misleading the reader is reduced to the minimum. Furthermore, the conclusions of this monograph are not entirely based upon the statistical material. Much collaborative historical evidence will be discussed in later chapters

[1] For bibliographical details of the gazetteers used in this chapter, the reader should refer to the Bibliography at the end of this book.

A STATISTICAL TABLE SHOWING THE HISTORICAL DEVELOPMENT AND GEOGRAPHICAL DISTRIBUTION OF WATER-CONTROL ACTIVITIES IN CHINA (Compiled according to data collected from provincial gazetteers)

DYNASTIC PERIODS / PROVINCES	Spring and Autumn (722–481 B.C.)	Warring States (481–255 B.C.)	Ch'in (255–206 B.C.)	Han (206 B.C.–A.D. 221)	Three Kingdoms (221–265)	Tsin (265–420)	Southern and Northern Dynasties (420–589)	Sui (589–618)	T'ang (618–907)	Five Dynasties (907–960)	Northern Sung (960–1127)	Southern Sung (1127–1280)	Sung (Miscellaneous Data)	Sung (Total for whole Dynastic Period)	Kin (1115-1260)	Yüan (1280–1368)	Ming (1368–1644)	Ch'ing (1644–1912)	Total Items for each Province	Date of Publication of Gazetteers
Shensi . .	..	..	1	**18**	2	..	..	9	**32**	**4**	12	**4**	**4**	20	**4**	12	48	38	208	1735
Honan . .	1	3	..	**19**	10	**4**	..	**4**	11	..	7	..	**4**	11	2	**4**	24	843*	947	1767
Shansi . .	1	..	..	**4**	1	1	1	3	**32**	..	25	..	..	25	14	29*	97*	156*	389	1734
Chihli (Hopei) .	..	..	..	5	1	2	3	1	**24**	..	20	..	..	20	**4**	**11**	**228**	**542**	886	1884
Kansu . .	..	..	..	1	1	..	..	..	**4**	..	2	..	..	2	..	2	19	19	50	1736
Szechwan . .	..	1	..	..	1	..	..	..	**15**	1	..	**4**	1	5	..	1	5	19	53	1815
Kiangsu . .	3	2	..	1	3	2	8	1	**18**	..	**43**	**74**	..	**117**	..	**28**	**234**	**62**	595	1736
Anhwei . .	1	..	..	1	3	..	4	1	12	..	7	9	..	16	..	2	30	41	127	1877
Chekiang . .	..	2	..	**4**	2	3	2	2	**44**	1	**86**	**185**	31	**302**	..	**87**	**480**	**175**	1406	1736
Kiangsi . .	..	..	..	1	..	1	1	..	20	1	18	**36**	2	**56**	..	13	**287**	**222**	658	1732
Fukien . .	..	..	..	..	..	2	..	**4**	**29**	..	**45**	**63**	294	**402**	..	**24**	**212**	**219**	1294	1754
Kwangtung. .	.. ..	..	..	..	..	..	..	..	..	..	16	**24**	**4**	**44**	..	**35**	**302**	**165**	536	1822
Hupei. . .	..	..	..	..	..	1	..	..	**4**	**4**	**4**	14	3	21	..	6	**143**	**528**	728	1921
Hunan . .	..	..	..	1	..	..	..	2	7	2	5	..	..	2	..	3	**51**	**183**	209	1885
Yünnan .	..	..	..	1	..	..	1	..	1	..	..	..	..	..	..	7	**110**	**292**	412	1736
Total Items for each Dynastic Period	6	8	1	56	24	16	20	27	254	13	290	543	363	1116	24	309	2270	3234		

to substantiate and elaborate the facts brought out in this chapter.

Source and Nature of the Data

This is the first time such a study has been attempted. The data used are collected and compiled from records of water-control activities listed in the chapters on water-control in the provincial gazetteers. The chapters are variously designated by such headings as "Rivers and Canals," "Water-benefit," or "Dikes and Dams." The construction or repair of any water-control work, be it a canal or embankment or a tank or any other kind of work, is regarded as an enterprise of water-control activity, and each record is put down in the table as one item.

Records of such activities, when undated or not clearly attributable to a dynastic period, were deliberately omitted, because, unless dated, at least by dynasties, this kind of record has no meaning as far as the present study is concerned.

The size and importance of the different works has to be disregarded, because, even when such data are available, they are not comparable. Fortunately the form in which the records are listed in the gazetteers takes care of the variation in size in the most desirable manner. The records are invariably listed by *hsiens*, or districts, and larger works, extending over several districts, are listed several times. Thus the larger the work the more frequently it appears in the records. This serves excellently the purpose of the statistical device of weighting, and gives proper valuation, in the data listed in the table, to the actual magnitude and importance of the activities.

That the provincial gazetteers are the best source for this kind of data will be clear when the process of compilation and the nature of the contents of such gazetteers are explained. The provincial gazetteers are compiled by appointed officials from the material in the local *hsien*, *chou* and *fu* gazetteers. A foreword to the Ch'ien Lung edition (1736) of the *Gazetteer of Kiangnan* (Kiangsu and Anhwei) remarks that the Emperor " commanded scholar-officials to prepare the Gazetteer of the United Ta-Ch'ing Domain (Ta-ch'ing-I-Tung Chih) and ordered the provinces by edict to compile provincial gazetteers to be submitted to the court. The responsible officials of the provinces in turn commandeered the gazetteers of the *chün* and *hsien* to be used as sources." [1]

To be sure, not all the material in the local gazetteers has been copied in the provincial gazetteers, but as far as the data here are concerned, little or practically no mutilation is discernable, and the more or less systematized presentation greatly lessens the labour involved in compiling the table which, even when confined to the data in the provincial gazetteers, is considerable. The ultimate sources from which the editors of the local gazetteers drew their material are the dynastic histories, government archives, inscriptions on stone tablets in which China abounds, writings of national, and especially local celebrities, and even traditions circulating as true history among the population. Relevant material from the rich literary heritage of China, the classics, history and literature was dug out and included in the gazetteers. Much local material, especially inscriptions on tablets, biographies of local celebrities, and geographical, economic and

[1] *The Gazetteer of Kiangnan.* Foreword by Yin Hui-yi, 1736, p. 1.

other data on the institutional life of the people, was thus published for the first time and the gazetteers are the only place to look for this material. The present study is only a humble beginning and purports to be one of the first few attempts to use this remarkable source material for the study of the economic and social history of China.

Interpretation of the Data

Before analyzing and interpreting the data in the table, it will be necessary to dispose of one point which critics of this volume are bound to raise at the outset. It refers to the question of the relation between climatic pulsation and the rise and fall of irrigation activities. After examining the work of various authorities on the climatic question in Asia, including that of Co-ching Chu, the British writer Brooks concludes that

> There is a general tendency for the number of floods to increase relatively to the number of droughts in the later centuries; when this is allowed for, the fourth, sixth, and seventh centuries and, later, the fifteenth and sixteenth centuries stand out as predominantly dry; the second and third, eighth, twelfth, and fourteenth centuries as wet. The general agreement with the results of the similar tabulation for Europe is very good.[1]

Comparing this paragraph with the table printed at the beginning of this chapter, it will be shown that no correlation between the two sets of facts can be established at all. If there is any such correlation, this table does not show it. All that this table shows is the relation between water-control activity and the political and economic importance of an area, which is the

[1] C. E. P. Brooks, *Climate Through the Ages*, London, 1926, p. 365.

problem that this monograph purports to study. Climatic pulsation may alter the degree of effectiveness of certain irrigation projects, but it cannot do away with the necessity for irrigation, because an average high level of rainfall does not necessarily lessen the rainfall variability from year to year, nor does it change the nature of the soil. Thus climatic pulsation can best be treated as a complicating factor and cannot seriously affect the main lines of the conclusion of the present study.

Looking at the table for the purpose of comparing water-control activities in the fourteen provinces during the various dynasties, it will be noticed that certain numbers (heavy type in the table) stand out among the rest and point to the following facts which, taken together, tell a story of immense importance :

(1) The table shows a number of water-control works constructed during the two periods of the spring and autumn (722-481 B.C.) and the Warring States (481-255 B.C.). Shansi records one work constructed between 676 and 652 B.C. In Anhwei the famous Shih Tank[1] was supposed to have been built by the Ch'u minister Sun Shu-ao between 606 and 584 B.C. In Honan another work ascribed to Sun Shu-ao and the famous works of Hsi-men Pao and Shih Ch'i were recorded.

In Kiangsu, the Han Kou, the earliest canal connecting the Hwai River with the Yangtze, was dug in this period.[2] A lake in Kiangsu and a tank in Chekiang were attributed to Fan Li and one work in each of the two provinces was attributed to Wu Yüan. Both Fan and Wu were outstanding personages in the period of the Warring States. Another canal in the city of Wu

[1] Cf. *infra*, p. 66-67. [2] Cf. *infra*, p. 65.

was supposed to have been constructed by Ch'un-shen-chün (Huang Hsieh) between 314 and 256 B.C. One creek in Kiangsu was named after T'ai Po, the eldest son of King Wen of Chou, and tradition has it that he dug it. In Honan, Chou Kung, a brother of King Wu of Chou, was supposed to have been the sponsor of one small irrigation canal.

It is difficult to ascertain the authenticity of these records. Except for the Shih Tank, the works of Hsi-men Pao and Shih Ch'i, and the Han Kou discussed in the following chapter, all of which are of great historical importance, the existence of the rest must be regarded as questionable and not worth serious consideration.

(2) During the Han dynasty (206 B.C. to A.D. 221) Shensi and Honan are credited with the largest number of items, the former having 18 and the latter 19. The next largest number, in Chihli, is only 5. The economic importance of the territory covered by the two provinces scarcely admits of doubt. Designated as Kuanchung and Honei respectively, they together constituted the Key Economic Area of the Han dynasty.

(3) During the period of the Three Kingdoms (190-264), the Tsin (265-419) and the Southern and Northern dynasties (317-586), there was an increase of water-control activities in the southern provinces. This is significant, especially in view of the fact that the number in the northern provinces declined.

Both Hupei and Fukien began to appear in the records for the first time, and the 8 items recorded for Kiangsu under the Northern and Southern dynasties, the largest number recorded for Kiangsu up to that time, is significant. It is also significant that Honan, Ts'ao Ts'ao's base during the period of the

Three Kingdoms, has a record of 10, while Anhwei, the battleground the development of which by Wei enabled it to conquer Wu, has a record of 3, the largest number for that province up to that period.

(4) During the T'ang dynasty (618-907) an unprecedented increase of water-control activities in practically all the provinces except Honan is to be noted, especially a sharp upward turn in all the southern provinces except Yünnan.

For the first time in Chinese history the number in a southern province, that of 44 in Chekiang, outstripped every province in the north. To be sure, Shensi, the seat of the capital, and Shansi, the " home " province of the dynasty, both received considerable attention, and each has a record of 32. But it is extremely significant that all of the provinces in the lower Yangtze valley, including Fukien, advanced above 18. Chekiang had a remarkable increase from 4 to 44, Kiangsi from 1 to 20 and Fukien from 4 to 29. Even Anhwei leaped from 4 to 12, Hunan from 2 to 7 and Hupei from 1 to 4.

Taking all the provinces together, it appears that the South at this period finally caught up with the North. The changing status of the South and the North, as well as the general sharp advance, indicates that the Key Economic Area had shifted, and marks the T'ang period definitely as the beginning of a new epoch in Chinese history.

(5) During the Sung dynasty (960-1280) the table shows another long step forward in the direction of water-control development in the Yangtze provinces, especially in Kiangsu, Chekiang, Kiangsi, and Fukien, where for the first time in Chinese history the number reached three digits, with Chekiang and Fukien leading,

the former having a record of 302 and the latter of 402. Hupei also advanced rapidly from 4 to 21. It is important to notice that Kwangtung entered the picture for the first time with 16 items in the Northern Sung and 24 in the Southern Sung.

The increase of water-control activity in these provinces was particularly noteworthy during the Southern Sung dynasty (1127-1280), which indicates that the settlement and development of the lower Yangtze valley, as well as that of the Pearl River valley, received a great impetus from the nomadic invasion of the north. Another important point to be noted is that Chekiang has 185 during the Southern Sung, an unusually large number. When it is recalled that the capital of the Southern Sung was situated at Hangchow, the meaning of this extraordinary growth needs no further elaboration.

(6) During the Yüan (1280-1368), Ming (1368-1644) and Ch'ing (1644-1912) dynasties, there are three features that deserve the closest attention. One is the growing importance of the Yangtze provinces, as well as Kwangtung, continuing the development from the T'ang and Sung dynasties and positively converting the Yangtze valley into the Key Economic Area. The second feature is the drastic increase in water-control activities during the Ming dynasty in Hupei, Hunan and Yünnan, which have respectively 143, 51 and 110 items listed, in contrast to 6, 3 and 7 during the Yüan dynasty.

The third feature is the special favour that was accorded Chihli during the three dynasties, while most of the other northern provinces suffered neglect. This is an unmistakable indication of a fear on the part of the authorities, repeatedly expressed in memorials to the throne, and engendered by the great distance of the Key Economic Area from the political base, which was responsible for a desire to develop the Hai Ho basin, in Chihli province, into a Key Economic Area, or, in the words of some officials, a "second Kiangnan."[1]

(7) It is necessary to explain separately the four starred numbers in the table. In the case of Honan, which has the extraordinary number of 843 during the Ch'ing dynasty, there is a defect in the way the activities are recorded in the provincial gazetteers. At least 90 per cent. of the items recorded are very small ditches or embankments, usually not affecting more than a village, built in the fifth, sixth and seventh years of the reign of Emperor Yung Cheng (1727-1729). The works involved are so small that the number has swelled to unbelievable proportions. Such data clearly belong to the category of exceptions.

In the case of Shansi, the numbers during the Yüan, Ming and Ch'ing dynasties, especially during the latter two, were swelled by private works. Looking over the gazetteers of all the eighteen provinces of China proper, Shansi seems to be the only province where private water-control works abound. This is probably the result of the high commercial development of that province in the last five or six centuries. It is clearly an exception to the data on the other provinces.

Although only fifteen provinces are included in the table, all the gazetteers of the eighteen provinces of China proper were examined. Kwangsi and Kweichow are missing because, surprisingly enough, the provincial gazetteers do not give the necessary data.

[1] This is the name given to the richest provinces south of Yangtze, Kiangsu and Anhwei.

As for Shantung, almost all water-control activities in that province were confined to flood-control on the Yellow River and the maintenance of the Shantung section of the Grand Canal. The data are given in such form as not to be amenable to any treatment that would make them comparable with the figures for the other provinces. Another weakness in the table is that the data for the Ch'ing or Manchu dynasty are not very adequate, since most of the gazetteers were published at different times before 1911, and therefore figures for the later years of the dynasty are incomplete. Fortunately the Ch'ing case is not as bad as it appears at the outset, because activity in public works declined during these later years as the dynasty itself weakened, and the data of the fifteen provinces included are sufficiently adequate to bring out the general line of the central thesis of this monograph.

CHAPTER IV

ORIGIN OF WATER-CONTROL AS AN ECONOMIC FUNCTION OF THE CHINESE STATE

THE origin of water-control works in China is traditionally linked with the historical fiction of the great Yü and the half-fiction of the Well-land System of ancient China. The cloud enveloping these two points must be cleared before it will be possible to emerge from the obscurity of twenty centuries of romance and mysticism and get a more or less clear view of the circumstances which, when collated with the few existing reliable direct evidences, would give us a general idea of the origin of water-control public works in China.

THE LEGEND OF YÜ AND THE DELUGE

Recent enthusiasm in the field of textual criticism of the classics by Chinese scholars has produced very interesting and illuminating results. However, so much work has been done in the field that there is danger of its degenerating to the level of a wild goose chase. Further research in this direction will not be very fruitful in results unless there is a new understanding of the controlling factors in the historical situation, or new factual discoveries through archæological research.

Pending confirmation from archæological excavations and allowing for new interpretative insights which may alter the conclusions without new evidence, the best conclusions of recent textual criticism can justifiably be treated as no more than valuable hypothesis. Except for these reservations the conclusions of Ku

Chieh-kang, the outstanding young exponent of textual criticism in China in regard to Yü, deserves the most serious consideration.

A daring analysis and comparison of the various sources of ancient Chinese historical literature led Ku to reject the traditional view about Yü and the deluge, which regards Yü as the great engineer-ruler who confined the wild-flowing rivers of China to their courses, delivered China (North China) from the Great Deluge, and founded the Hsia dynasty.[1] Ku advanced the theory that the traditional version of the ancient history of China was built up in successive strata by writers in different epochs and that the accepted order of the supposedly historical personalities and events was exactly the reverse of their true historical order.

Ku thinks that "Yü . . . made his appearance in the Western Chou period (1122-770 B.C.), but that Yao and Shun first emerged in the later years of the spring and autumn era (722-481 B.C.). That is to say, the earlier these figures appear in the accepted chronology, the later they actually emerged in history. As a matter of fact, Yü was known before Yao and Shun; and Yao and Shun were known long before Fu-hsi and Shen-nung, but in traditional chronology the order was entirely reversed."[2]

In the case of Yü, Ku's version is that he was a god in the mythology of the people of the Yangtze valley about the time between the Yin and Chou dynasties, about the eleventh century B.C. The legend must have centred around Kw'aichi which is now called

[1] The whole discussion between Ku and various other scholars on the question can be found in Ku Chieh-kang's *Discussions in Ancient Chinese History*, Peiping, 1930, vol. i, especially pp. 59-142 and pp. 206-210.

[2] Arthur W. Hummel, *The Autobiography of a Chinese Historian*, Leyden, 1931, p. 97-98.

Shaohsing, in Chekiang province. The people of Yüeh worshipped Yü as their ancestor and his grave is supposed to be in Kw'aichi. From Kw'aichi the legend travelled to T'ushan in Anhwei province where Yü was supposed to have summoned the chiefs of various tribes to meet him. From T'ushan it went to Ch'u (present Hupei province) and from Ch'u to North China. The transmission of the legend to North China is largely the result of frequent contact of the northern powers with Ch'u, through war or otherwise, since the time of King Chao of Chou (traditionally 1026-1002 B.C.). This is perhaps the reason why only since King Mu of Chou (traditionally 1001-947 B.C.) was Yü's name mentioned in the most authentic records of the history of North China, the *Book of Poetry* and the reliable portions of the *Book of History*. Thus it seems that the periphery of Yü's activities and influence increased with the spread of the legend, and became universal when in the latter part of the Chou dynasty it reached the heart of Chinese civilization in the Yellow River valley.

Ku believes that it was the peculiar geographical conditions of the Yangtze valley, the menace of forests, wild beasts and swamps, the great scourge of flood, especially in the valley of the Ch'ient'ang River (then a tributary of the Yangtze), and the consequent pressing need of water-control that gave rise to the legend of Yü and the Deluge.

Mencius' account of the legend states that :

> In the time of Yao, when the world had not yet been perfectly reduced to order, the vast waters, flowing out of their channels, made a universal inundation. Vegetation was luxuriant, and birds and beasts swarmed, the various kinds of grain could not be grown. The birds and beasts pressed upon men. The paths marked by the feet of beasts and the prints

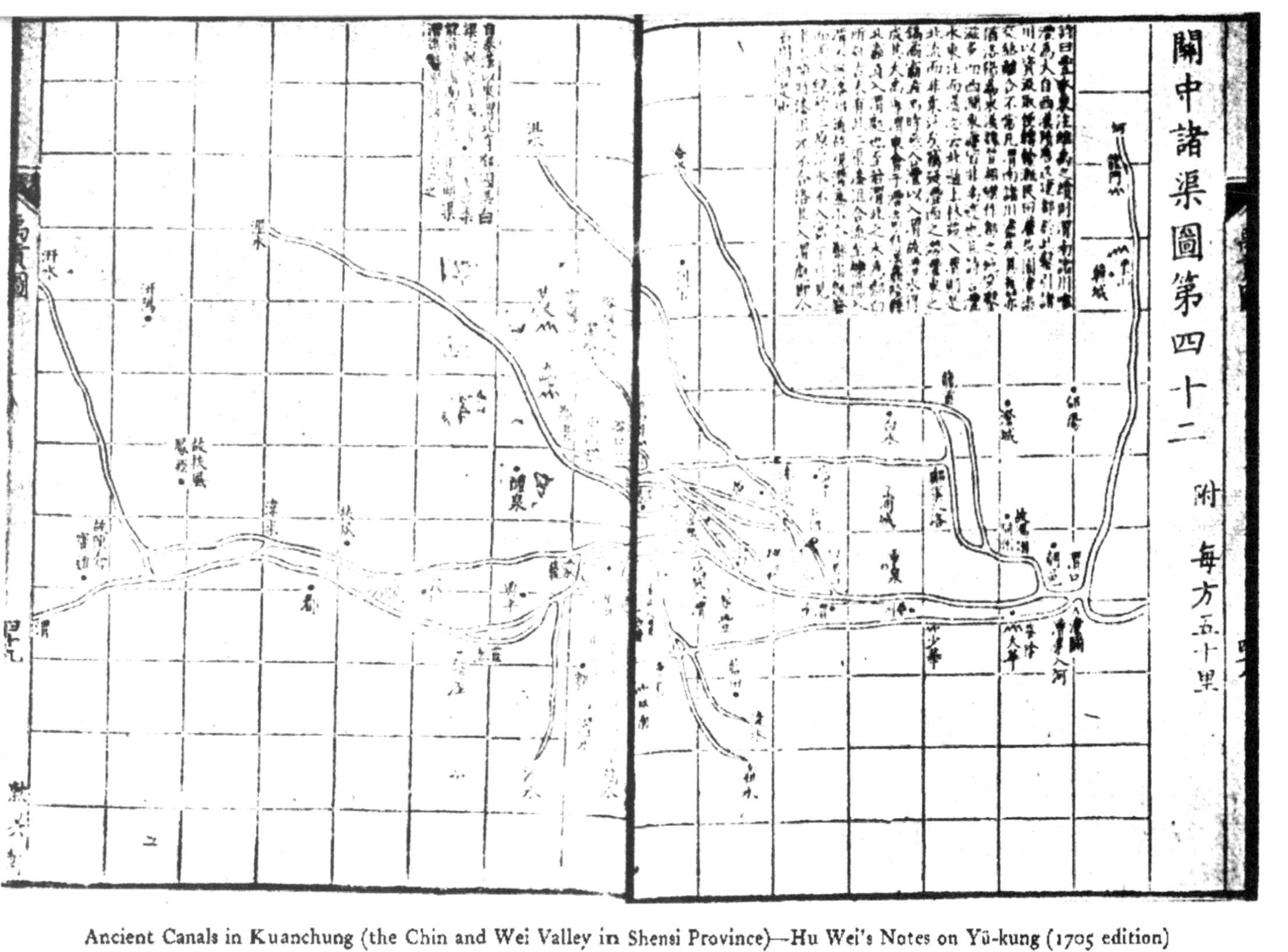

Ancient Canals in Kuanchung (the Chin and Wei Valley in Shensi Province)—Hu Wei's Notes on Yü-kung (1705 edition)

of birds crossed one another throughout the Middle Kingdom. . . . Yao raised Shun to office, and measures to regulate the disorder were set forth. Shun committed to Yü the direction of the fire to be employed, and Yi set fire to, and consumed, the forests and vegetation on the mountains and in the marshes so that the birds and beasts fled away to hide themselves. Yü separated the nine streams, cleared the courses of the Tsi and Ta, and led them all to the sea. He opened a vent for the Ju and Han, and regulated the course of the Hwai and Sze, so that they all flow into the Chiang. When this was done, it became possible for the people of the Middle Kingdom to cultivate the ground and get food for themselves.[1]

In another place Mencius describes ancient China thus:

In the time of Yao, the waters, flowing out of their channels, inundated the Middle Kingdom. Snakes and dragons occupied it, and the people had no place where they could settle themselves. In the low grounds they made nests for themselves on the trees or raised platforms, and in the high grounds they made caves. It is said in the Book of History, "The waters in their wild course" were the waters of the great inundation. Shun employed Yü to reduce the waters to order. Yü dug open their obstructed channels, and conducted them to the sea. He drove away the snakes and dragons, and forced them into the grassy marshes. . . .[2]

Ku suspects that Mencius' account of the geographical conditions of ancient China is a description of the conditions of Ch'u (Hupei) and Yüeh (Chekiang) in his time, which Mencius thought might have been similar to the ancient conditions of North China. There can be little doubt about the fitness of this description for the Ch'u and Yüeh of Mencius' time, especially the southern part of Chu, which had not yet been opened up. On the other hand, modern geological knowledge tells us that the part of ancient China which was situated in the loess steppes could have had neither

[1] James Legge, *The Work of Mencius*, Oxford, 1895, vol. ii, pp. 250-251.
[2] *Ibid.*, pp. 279-280.

dense forests and heavy vegetation, nor dangerous marshes; neither could the part which was situated in the alluvial plain of North China have had the dense forests and heavy vegetation which Mencius so eloquently described.[1] Thus, according to Ku, the legend of Yü and the Deluge most likely came to North China from the Yangtze River valley. This conclusion about the origin of the Yü legend, however, does not imply that the Yangtze valley was developed before the Huang Ho valley, and does not alter the fact that the latter and not the former region is the cradle of Chinese civilization.

Whether the positive contributions of Ku's conclusions will be supported or refuted by newly discovered evidence in the future, his scathing criticism of the traditional version of the legend seems to have successfully destroyed the mystical theory which attributed the beginning of public works for water-control in China to the benevolent and self-sacrificing activities of a kingly hero. Through centuries of repeated affirmation by orthodox scholars this mystical theory has acquired the authority of a religious faith and has become a great obstacle to any scientific study of the problem. With its complete demolition, an objective study of the available data on the beginnings of water-control activities thus becomes possible.

Earliest Record of Irrigation Practices and the Well-land System

The first authentic record of irrigation practice in China is to be found in the Shih Ching,

[1] See V. K. Ting, "Professor Granet's 'La Civilization Chinoise,'" *The Chinese Social and Political Science Review*, vol. xv, No. 2, Peiping, July 1931.

or *Book of Poetry*, which contains the following verse :

> How the water from the Piao pool flows away to the north,
> Flooding the rice fields ! [1]

The poem was supposed to have been written by the Queen of King Yü (781-771 B.C.) of the Chou dynasty. It is significant that the first work of irrigation mentioned in the records is located in the heart of the loess region in central Shensi. The *History of the Warring States*, which was the work of writers who lived about the middle of the first millennium before Christ, also refers to irrigation and tells us that when, about the fifth century B.C., Eastern Chou wanted to cultivate rice, its enemy, the Western Chou, which was situated upstream along the Yellow and Lo Rivers and therefore was in control of irrigation canals connected with the two rivers, refused to release the water, and Eastern Chou was forced to cultivate wheat.[2] Though brief, these records are reliable and they clearly show that irrigation was practised in China at the eighth and fifth centuries before Christ. Further back than this, no record is yet available.

It is very significant to note that, at the beginning, irrigation must have been practised on a very small and local scale. The record in the *Book of Poetry* merely refers to a pool, while irrigation canals in the *History*

[1] James Legge, *The Chinese Classics*, vol. iv, part II, p. 416. The translation follows Legge except for the sixth and seventh words in the first line. Here Dr Legge, following older authorities, just translates "pools," treating the word Piao as an adjective describing the appearance of flowing. Modern Chinese scholars think that "Piao pool" is the name of a small lake which once existed in the western part of Hsian district, Shensi province. See *Anthology of Ancient Chinese Studies* (*Book of Poetry*), with Introduction and Notes, by Min T'ien-shou, Shanghai, The Commercial Press, 1928, p. 26.

[2] *History of the Warring States*, selected and annotated by Tsang Li-ho, Commercial Press, Student Edition, Shanghai, 1932, pp. 3-4.

of the Warring States could not be of any considerable size. Confucius, who lived in the sixth century B.C., referred to ditches and furrows, but gave no indication of the existence of large irrigation works in his time. Chinese scholars, following Confucius, always refer to the irrigation system of ancient China as the "System of Ditches and Furrows," which is regarded as an integral part of the "Well-land System" of ancient China, particularly of the Chou period. The true nature of the traditional irrigation system cannot be grasped without a proper understanding of the "Well-land System."

The "Well-land System" has been a subject of bitter controversy ever since it was first described by Mencius. In order to obtain a clear idea of the original version of the system, it will be necessary to quote Mencius in full. Mencius gave an outline of the system in answer to a question by Pi Chan, who was sent by the Duke of T'eng to ask Mencius about the "Well-land System." Mencius says:

> Since your prince, wishing to put in practice a benevolent government, has made choice of you and put you into this employment, you must exert yourself to the utmost. Benevolent government must begin with the land boundaries. If the land boundaries be not defined correctly, the division of the land into the Well System will not be equal, and the produce available for salaries will not be equitably[1] distributed. On this account, oppressive rulers and impure ministers are sure to neglect this defining of the boundaries. When the boundaries have been defined correctly, the division of the fields and the regulation of allowances may be determined by you, sitting at your ease.
>
> Although the territory of Tang is narrow and small, yet there must be in it men of a superior grade, and there must

[1] Legge translates this "evenly," but the original word *ping* contains the idea of justice and does not mean an equal distribution in this case. It can be better rendered by the word "equitably."

be in it country men. If there were not men of a superior grade, there would be none to rule the country men. If there were not country men, there would be none to support the men of superior grade.

I would ask you in the remoter districts, observing the nine-square division to reserve one division to be cultivated for the lord by the eight peasant families, who are given the remaining eight squares for themselves, and in the more central parts of the kingdom, to make the people pay a tax of a tenth part of their produce.

From the highest officers down to the lowest, each one must have his holy field, consisting of fifty *mu*.

Let the supernumerary males have their twenty-five *mu*.

On occasions of death, or removal from one place to another, there will be no quitting the district. The fields of a district are all within the Well unit. The peasants attached to this unit of land render friendly offices to one another in their going out and coming in, aid one another in keeping watch and ward, and sustain one another in sickness. Thus the people are brought to live in affection and harmony.

A square *li* covers nine squares of land, which nine squares contain nine hundred *mu*. The central figure is the public field, and eight families, each having its private hundred *mu*, cultivate in common the public field. And not till the public work is finished, may they presume to attend to their private affairs. This is the way by which the country men are distinguished from those of a superior grade.

Those are the great outlines of the system. Happily to modify and adapt it depends on the prince and you.[1]

The above outline was given by Mencius (372-289 B.C.) more than two centuries after the first important legal act signifying the breakdown of the feudal system

[1] James Legge, *The Works of Mencius*, vol. ii, pp. 243-245. The above quotation is not the exact wording of Legge's translation. It is a corrected version based upon his text. The most important mistake Legge made in his translation is in the third paragraph, his version being as follows: "I would ask you in the remoter districts, observing the nine-square division to reserve one division to be cultivated on the system of mutual aid, and in the more central parts of the kingdom, to make the people pay for themselves a tenth part of their produce." He evidently misunderstood the system of *Chu* as a system of mutual aid, while in reality it was a system of taxation, being further explained in the seventh paragraph. I have substituted, for the Legge version, a phrase explaining the system which in the Chinese text is only indicated by the one word *Chu*.

of landholding was recorded in history.[1] He did not indicate his sources except in another connection, when he cited a verse from the *Book of Poetry*. It reads:

> May the rain come down on your public field,
> And then upon our private fields.[2]

New Interpretation of Mencius

Traditional Chinese scholars who regard Mencius as correct in every detail have evidently placed their loyalty to the master above considerations of historical scholarship. Nevertheless careful scholarship does not preclude viewing Mencius' work as a valuable clue.

The geographical conditions of North China are such that it would be natural for wells, or a deep pit with a sufficient supply of underground water, to become a central feature of the land system in a locality most favourable to its development. The well-land, or *Ching-t'ien*, could have been a manorial administrative unit, with the well as its nucleus. Kuo Mo-jo tells us that, according to the inscriptions on bronze of the Western Chou period (1050-781 B.C.), there was a feudal state named Ching (Well) near Tasan Kuan (the present Paochi district, Shensi Province).[3] Kuo did not see the connection, but it is very likely that the Ching state was so called because of its well-land system; which name might have gradually been

[1] The *Spring and Autumn* records that in 594 B.C. the state of Lu began to levy a land tax according to *mu* or acreage. This signifies the decline of the feudal land system of making the serfs contribute their labour to work on a piece of the lord's land, the produce of which would be handed over exclusively to the lord. The land tax was probably made necessary by the breakdown of feudal landholding and the rise of the institution of private land ownership.

[2] *Ibid.*, vol. ii, p. 242.

[3] Kuo Mo-jo, *A Study of Ancient Chinese Society*, Shanghai, 1931, p. 300.

extended to designate the land system of the whole epoch in general.

The form of the word Ching, 井, might have suggested the idea to Mencius that the unit of the ancient manor was actually divided into nine squares. Later writers, following Mencius, added more and more details in trying to elaborate his point until a very artificially worked-out ideal land system was constructed in the *Chou Li* or *Institution of Chou*, which the weight of modern authority considers to have been written at the end of the Earlier Han Dynasty (206 B.C. to A.D. 25). At the same time the seigniorial aspect of the system was less and less noticed by these later writers, until finally it was forgotten altogether.[1] Both the apparent absurdity of the artificial form of nine squares and the obvious contradiction of a feudal society with an idyllic communist system of land-holding induces many sceptical writers to deny the existence of a Well-land System altogether and denounce the whole thing as purely a mental fabrication.[2]

If, however, the question of artificial regularity is dismissed as an unimportant detail, and Mencius' outline of the system interpreted in the light of the purpose for which he proposed its readoption, and the corroborative evidence in regard to the general socio-economic conditions of the time, it will be found that the system can be reconstructed which is not only plausible, but reveals, in clear outline, the essential features of the basic productive socio-economic unit in

[1] See Liao Chung-k'ai's letter to Hu Shih, printed in *Essays of Hu Shih*, Shanghai, 1921, vol. i, pp. 253-263.

[2] See the very interesting discussion of the problem by Dr Hu Shih in his letter to Liao Chung-k'ai and Hu Han-min in *Essays of Hu Shih*, vol. i, pp. 264-282. The letter contains the germs of many interesting ideas. But he made the mistake of considering Mencius' statement to be purely a result of his imagination and without any historical basis.

ancient China, the primitive village community, which by the time of Mencius was evidently going through a process of disintegration and differentiation.

The main purpose for which Mencius proposed the readoption of the system was to " define correctly the boundaries," thus bringing about "an equal division of the land," which would insure an " equitable distribution of the produce available for *lu*, or salaries." *Lu*, or " salaries," refers to the income of the feudal lord, namely, the whole produce of the " public " or seigniorial strip of a unit of well-land ; the " public " field being jointly cultivated by the serf families, who also cultivated their own " private " portions on a subsistence basis. Had the productivity of labour remained on a stagnated level, the size of the well-land unit would generally be measured by the productive capacity of the peasants, and the size of the " private field " for the support of each peasant-serf family would most likely be limited by the customary cost of subsistence. The land profitably held by the feudal lord would thus be determined by the number of peasant-serf families under his control. Feudal wars for serfs and land thus would have been the order of the day; but internally, the social system probably would have remained stable so long as the productivity of labour did not undergo any significant change.

But with the introduction of iron into North China in the middle of the first millennium before the Christian era,[1] the beginning of the use of oxen for

[1] H. T. Chang, "Studies in the Development of Bronze and Iron Age in China," *Memoirs of the Geological Survey of China*. The manuscript of the English translation of this monograph, by Dr Arthur W. Hummel, is in the Library of Congress, Washington, D.C. Also see C. W. Bishop, " The Geographical Factor in the Development of Chinese Civilization," *Geographical Review*, vol. xii, New York, January 1922, p. 32, and C. W. Bishop, " Rise of Civilization in China," *Geographical Review*, vol. xxii, No. 4, October 1932, pp. 630–631.

ploughing at about the same time,[1] and the increasing use of animal manure as fertilizer,[2] the productivity of the Chinese peasant-serf must have been sharply augmented. The lord naturally saw in this change a splendid opportunity for him to reduce the size of the "private field," thus appropriating more of the peasant's labour to his benefit.

The attempt by "oppressive rulers" and their "impure ministers" to reap the fruit of this tremendous revolution in agricultural productivity by changing the boundaries between the public and "private" fields could hardly escape serious social consequences. The situation was made worse by the fact that a serf who was out of favour might suffer a larger inroad on his share of land than others. Mencius saw in this excessive and unequal exploitation a cause for social unrest. There must have been a tendency for peasant-serfs to attempt to escape increased burdens by leaving the land and taking to other occupations which the development of handicraft industry and commerce of the time offered. He proposed the restoration of the Well-land System as a means of solving the socio-economic problem of his time by lightening and equalizing the burden of the peasant-serfs through equal division of land, which, he thought, could be done by a clear defining of land boundaries.

The so-called "private field," allotted to the serf, is by no means to be considered private property in the modern sense of the word. Like the medieval serfs in Europe the ancient Chinese serfs merely held the land in use in exchange for certain services to the lord,

[1] Hsü Chung-shu, "On Some Agricultural Implements of the Ancient China," *Academia Sinica, Bulletin of the Natural Research Institute of History and Philology*, Peiping, 1930, vol. ii, part I, p. 58.

[2] Wan Kuoh-ting, *An Agrarian History of China*, Nanking, 1933, p. 37.

primarily labour on the public fields and military service. The "public field" resembles very much the demesne in English medieval history. Both the "private" and the public fields were really the property of the lord or "man of superior grade," as Mencius designates those who not only owned the land or held it from an overlord as a gift with hereditary rights, but also governed the peasant-serfs whom Mencius calls the "country men," and whose duty it was to "support the men of superior grade."

The well-land, like the manor in European history, was evidently the lowest administrative, as well as economic, unit in the feudal hierarchy. Even "on occasions of death or removal from one place to another there will be no quitting the district." The peasant-serfs were thus attached to the land. "The fields of a district are all within the Well unit. The peasants attached to this unit of land render friendly offices to one another in keeping watch and ward, and sustain one another in sickness." Thus the well-lands were well organized, economically self-sufficient units of agrarian organization, a kind of village community, the cells that constituted the basic structure of feudal society.

The number of well-land units either directly or indirectly under the jurisdiction of a feudal lord depended upon the position of the lord in the feudal hierarchy. Mencius, in answer to the question by Pei-kung I, who asked about the Chou system of feudal grading of dignities, gives a long list of such gradings and their respective landholdings, although he reveals that the details of the arrangement could not be learned, because the feudal lords of his time, in the period of the decline of feudalism, regarded the records as injurious to their interests and destroyed them.[1]

According to the list, " to the son of Heaven there was allotted a territory of a thousand *li* square ; a Kung (Duke) and a Hou (Marquis) had each a hundred *li* square ; a Po (Count) had seventy *li*, and a Tzu (Baron) and a Nan (Squire) had each fifty *li*. . . . The chief ministers of the Son of Heaven received an amount of territory equal to that of a Hou ; a great officer received as much as a Po ; and a scholar of the first class as much as a Tzu or a Nan. . . . As to those who tilled the fields, each husbandman (serf) received a hundred *mu*. . . ."

If, as stated in the first quotation,[2] " a square *li* covers nine squares of land, which nine squares contain nine hundred *mu*," the Son of Heaven according to the above list must have been allotted a thousand well-land units, with about eight thousand families under his direct jurisdiction. He, of course, governed the territory through his ministers, to whom a certain number of well-land units of land were allotted according to their grade. Besides the territory under the direct jurisdiction of the Son of Heaven, which was called *wang chi*, there were feudal states of various orders of size and importance, all nominally under the suzerainty of the Son of Heaven, but in practice autonomously governed by their own rulers and their own series of feudal subordinates of different grades, who were allotted a number of well-land units appropriate to their status in the feudal order. At the bottom of this elaborate feudal superstructure lay its economic base, the well-land units, each of which was

[1] James Legge, op. cit., vol. ii, pp. 373-376.
[2] Cf. *supra*, pp. 52-53.

a self-sufficient rural community, suffering under an ever-increasing burden of feudal exploitation.

To accept the statement of Mencius as a rough approximation of the system of feudalism in ancient China does not necessarily imply an acceptance of the traditional conception of the King of Chou as the supreme ruler of a single state covering the territory of the whole of China. The best interpretation of the remarks of Mencius on the question would be that the King of Chou exercised a sort of loose suzerainty over a number of feudal states scattered throughout the Huang Ho or Yellow River valley.

Pending the discovery of new material, through archæological excavation or otherwise, the best account of the land system of ancient China is to be found in the above quoted sections of Mencius; and the interpretation given here seems to be the most reasonable construction of the text in the light of modern knowledge of the economic history of the period.

The "Ditches and Furrows"

Thus, the well-land was a unit of rural economy and administration which formed the basis of Chinese feudal society. The "ditches and furrows," according to the traditional view held by most Chinese scholars, constituted the boundaries between the well-land units and the nine-square subdivisions of fields within each unit. The most detailed version of the system, obviously very much idealized and in a form too perfect and regular to be true, is described in the *Chou Li* about the beginning of the Christian era. Even a semblance of the system of ditches and furrows as described in the *Chou Li* could have existed only in South

China. It is very unlikely that all the boundaries of the well-land units throughout the territory of ancient China in the Yellow River valley could have been ditches and furrows. The *Chou Li* version of the system is largely an imaginary construction by later scholars, based upon fragmentary historical information and the observation of conditions in the Yangtze valley. There can hardly be any doubt, however, that in ancient China, at the time of Confucius or earlier, wells and ditches and furrows were historically in evidence; otherwise the Well-land System would not have been so called, and Confucius would not have had any cause to mention ditches and furrows. But these ditches and furrows must have been on a local scale, and perhaps no enterprise beyond the ability of the collective effort of the peasant-serf of a well-land unit to construct and maintain could have existed.

Social Prerequisites of Large-Scale Water-Control Development

None of the references to irrigation in ancient literature cited earlier in this chapter,[1] mentions canals or other irrigation works on a large scale. That such works evidently did not exist in classical feudal times is not difficult to understand. The socio-economic organization of feudal society was such that, as the above study of the ancient land system shows, all peasant-serf families were attached to the well-land unit and, unless the system broke down, surplus and unattached labourers could not amount to any significant numbers. Large mobilization of labour, under the circumstances, would necessarily have

[1] Cf. *supra*, pp. 50-51.

interfered with the regular routine on the farms, the effect of which would be immediately felt by the lords.

This is decidedly in contrast with the conditions in later times, especially since the Han dynasty, when there were literally hundreds of thousands of wandering labourers roaming the country, and when the system of private land ownership, with a more or less fixed minimum tax burden, obscured the danger of mobilization of the peasants for public works.

Prior to the appearance of these conditions, which made the mobilization of a large number of labourers a comparatively easy and not obviously objectionable undertaking, collective labour must have been mostly a local affair confined to the well-land units. Larger enterprises would certainly have been improbable, if not impossible, in classical feudal China. Besides, it is also likely that the inefficient tools and low engineering skill of the period which preceded the introduction of iron tools prevented the construction of canals or tanks of any considerable size.

The period of Warring States (481-255 B.C.) witnessed a tremendous technical as well as social revolution that finally ushered China into the ensuing epoch, the semi-feudal or imperial period of its history. As shown earlier in the chapter,[1] the beginning of the iron age, the use of oxen to pull the plough, and the increasing application of animal fertilizer to cultivation and the consequent revolutionary growth in the productivity of labour in agriculture, played havoc with the ancient land system and gradually brought about private land ownership.

This can be dated by the beginning of the institution

[1] Cf. *supra*, p. 56.

of a land tax levied according to *mu*,[1] irrespective of the public or "private" nature of the land and the identity of the actual labourer who worked on the land, in 594 B.C., in the State of Lu.[2] The increase in the productivity of labour and the breakdown of the Well-land System swept a large number of peasant-serfs off the land. At the same time the abolition of the difference between public and "private" fields and the assumption by private landowners, either minor lords or merchant princes, of a land tax assessed according to *mu* or acreage in lieu of the former direct contribution of labour on the public field, cut the string which tied the feudal lords to the routine of production and freed them from concern over the harmful effects of protracted and large-scale forced labour on agricultural production. Thus the revolution in the land-system created the conditions for large-scale mobilization of forced labour, and made possible the construction of large-scale public works for water-control.

Water-control Works as Weapons in Feudal Struggles

The gathering storm of the socio-economic crisis intensified the conflicts between the feudal states. It is interesting to note that the feudal authorities of the various states were quick to use the construction of water-control public works as a valuable new instrument to further their own interests in struggles with neighbouring states.

[1] A unit of land measurement approximately one-sixth of an acre. Legge's transliteration is "mau."

[2] Recorded in the *Spring and Autumn Annals*. Cf. *supra*, p. 54, note.

The famous Han authority on water-control, Chia Jang, tells us that "the building of dikes recently began during the Period of the Warring States when the various states blocked the hundred streams for their own benefit. Ch'i, Chao and Wei all bordered the Huang Ho. The frontiers of Chao and Wei rested on the foot of the mountains while that of Ch'i was on the low plain. Hence Ch'i constructed an embankment twenty-five *li* from the river, so that when rising water approached the Ch'i embankment, it would be forced to flood Chao and Wei. Hence Chao and Wei also constructed an embankment twenty-five *li* from the river [to counteract the effect of the Chi embankment]." [1]

In his book *Words about Ancient Things*, the Ming scholar, Cheng Hsiao (1499-1566), made the following revealing comment of the water-control activities of the period. He says:

> The decline of Chou was accompanied by the gradual abandonment of the Well-land system. Those feudal states which quarrelled for water-benefits, constructed dikes to enclose the fertile valleys and river channels for their own benefit. Those who were anxious to avoid the dangers of flood also constructed dikes to force water into their neighbour's country, regarding the latter as a reservoir for surplus water. [Thus] more and more dikes were built day by day and they encroached so much upon the natural channel of the river that the dikes were burst and floods became frequent.[2]

Both Chia Jang and Cheng Hsiao were referring to feudal struggles with water-works on the Yellow River as weapons, struggles to appropriate fertile river beds

[1] Pan Ku, *Earlier Han History*, *chüan* 29, Book on Canals and Ditches, p. 13.

[2] Quoted in Fu Tse-hung's *Golden Mirror of the Course of Rivers and Canals*, 1725, *chüan* 3, p. 5.

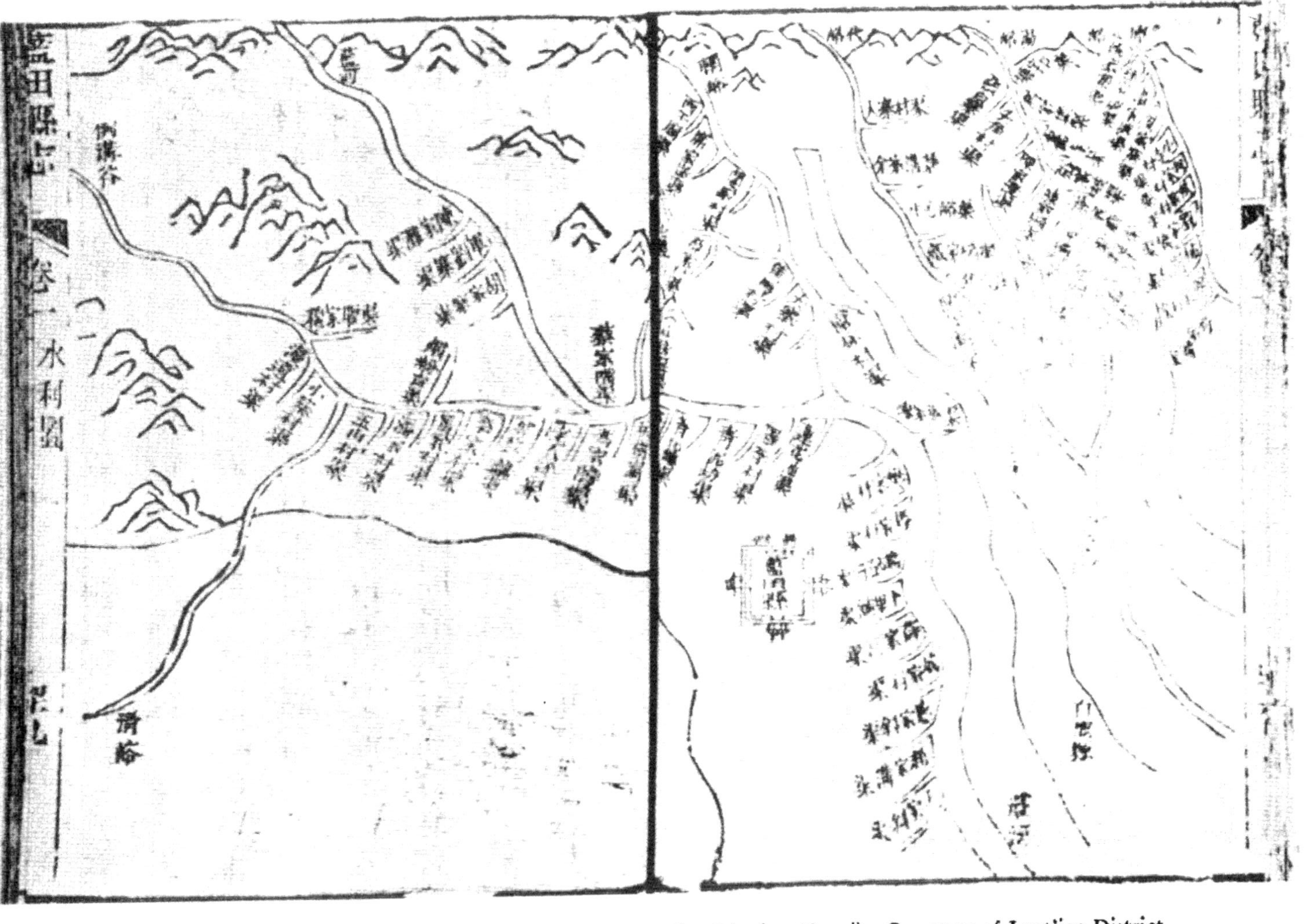

Typical Inundation Canal System in North China (Lant'ien District, Shensi)—Gazetteer of Lant'ien District

for cultivation and attempts to throw the menace of flood on one's neighbour.[1]

ARTIFICIAL WATERWAYS IN FEUDAL CHINA

However, the most significant use of water-control works as weapons in feudal struggles was the construction of artificial waterways for transportation. As the late-feudal wars became more and more frequent, rivers were widened and deepened and canals were dug for the purpose of transporting troops and especially provisions. The revolution in the land system, the levying of the land tax by the feudal princes and the development of commerce broke down the isolation of local feudal communities, and concentrated more and more power into the hands of the heads of the feudal states and necessitated the transport of tax grain or grain tribute to the seats of power. This, and the demands of commerce, also stimulated the construction of artificial waterways.

In 486 B.C., King Fu Ch'ai of Wu dug the Han Kou, the earliest canal connecting the Hwai River with the Yangtze, in order to facilitate his designs for northern conquest.[2] Ssu-ma Ch'ien (b. 145) speaks of the opening of the canal of Hung Ko (later called the Pien River) to connect the Yellow River south-eastward with the feudal states of Sung, Cheng, Ch'en, Ts'ai, Ts'ao and Wei, and link it with the four rivers, Ch'i,

[1] Irrespective of the primary aim of the works at the time of construction and the damage done by their competitive rather than co-operative nature, the objective result was improvement in the technique of constructing large water-control works gained through new experience. Once the technique was incorporated into the consciousness of a people, especially in China where water-control was of so much fundamental significance, it assumed a role in history the importance of which can hardly be exaggerated.

[2] *Tso Chüan*, Book of Ai-Kung, p. 22.

Ju, Hwai and Ssu, all of which led to the country of Ch'u. Canals were dug in the west to reach the Han River and the country of Yünmeng; on the lower Yangtze, in Wu, to reach the three *chiangs* (rivers)[1] and the five *hus* (lakes) (the present T'ai Hu); and in Ch'i to connect the Ch'i and Chi Rivers in Shantung province. In Shu (Szechwan province), the governor, Li Ping, cut a canal through Ch'engtu to connect the navigable rivers in the region into one system.

Ssu-ma Ch'ien significantly adds, after enumerating all these early efforts in canal building, that "all the canals were used for boats; and if there was sufficient water, they would then be used for irrigation. Peasants enjoyed the benefit; wherever the canals passed, the peasants made use of the water. There were tens and hundreds of thousands of ditches—nay, an incalculable number of them—to lead the water from the canals to the fields."[2]

FIRST LARGE-SCALE IRRIGATION WORKS

The earliest known public work constructed on a large scale primarily for irrigation purposes was the Shih Pei, or the Shih Reservoir, near the district of Shou in northern Anhwei province, which was supposed to have been built by the governor of Ch'u

[1] Traditionally considered as the three branches into which the lower course of the Yangtze in Kiangsu was divided.

[2] Ssu-ma Ch'ien, *Historical Records*, *chüan* 29, Book on Rivers and Canals, pp. 2-3. According to Ts'ui Shih, the whole book on rivers and canals in the *Historical Records* was not the work of Ssu-ma Chien, but was copied by later scholars from the Book on Canals and Ditches, a section in the *Earlier Han History*. See Ts'ui Shih, *Shih Chi T'an-yuan*, or *Researches into the Source of the Historical Records*, 2 vols, 8 *chüan*, published 1922 by National Peking University Press. *Chüan* 4, p. 18. However, as far as concerns the part upon which the material for this paragraph is based, the two versions are practically identical, and for our purposes, it really does not matter which author did the copying.

Sun Shu-Ao during the reign of King Ting of Chou (606-586 B.C.). Irrigating 40,000 *ching* of land, it was the most important irrigation project in the region for centuries, and was repeatedly repaired after the Han and Tang dynasties.[1]

The honour of being the second builder of a large and significant work primarily and specifically for purposes of irrigation must be assigned to Hsi-men Pao, a minister of Marquis Wen (403-387 B.C.) of the State of Wei. Ssu-ma Ch'ien relates the tale that, upon appointment as administrator of Po, a province of Wei, Hsi-men Pao discouraged the superstition of the people about a bride for the god of the river and punished the local gentry and bureaucrats who took advantage of such superstitions, and, in conspiracy with an old witch, extorted money and drowned in the river specially selected peasant maidens on the pretence that they had to arrange marriages for the river-god.[2] The superstition about a marriage to the river-god and the human sacrifice connected with it seems to have been common at the time. An old book, *Feng Su T'ung*, speaks of a similar practice in Szechwan, and relates the heroic efforts of the governor, Li Ping, to combat such practices in connection with his work in cutting irrigation canals and converting the peasants to the advantages of irrigation.[3] After he had succeeded in stopping such practices by a clever trick, he mobilized the peasants to cut twelve canals to carry water from the Yellow River to irrigate the people's land.

[1] K'ang Chi-t'ien, *Notes on Canals and Rivers*, 1804, *chüan* 2, p. 18.

[2] Po was situated on the north bank of the Yellow River in southern Hopei and northern Honan.

[3] See a note on the incident, as recorded in the *Historical Records*, *chüan* 29, Book on Rivers and Canals, p. 2.

Ssu-ma Ch'ien, who recorded the story, has this to say about the people's reaction to the project: "The fields were irrigated, but, at that time, the people were unwilling to suffer even slight trouble to work on the canals. Pao's comment was that, 'people can be depended upon to enjoy the results, but they should not be consulted about the beginning [of the task]. Now the elder ones and their descendants dislike people like me, but hundreds of years later let them think of what I have said [and done about irrigation].' "[1]

The dissatisfaction over forced labour (corvee) must have been the main cause of the people's complaint against Hsi-men Pao. It is likely that the hardships of forced labour were made even harder for the people to endure by the fact that, at this time, irrigation construction on a large scale was still an innovation, and the people were sceptical about its beneficial results.

The next authentic record of large-scale irrigation construction is a mention of the work of Shih Ch'i, who served under King Hsiang of Wei (318-296 B.C.), the great-grandson of the ruler under whom Hsi-men Pao worked. Shih Ch'i criticized Hsi-men Pao, before the king, for not utilizing the river Chang in Honan province for irrigation. He said that, while in other parts of Wei each peasant was given 100 *mu* of land to cultivate, in Yeh he was given 200 *mu*. This, he explained, was because the land was bad and must be enriched by making use of the river Chang. Having been made administrator of Yeh by the King, he canalized the region and harnessed the river Chang for

[1] Ssu-ma Ch'ien, *Historical Records*, *chüan* 126. Biographies of Humorists, pp. 12-15.

irrigation to enrich the Honei district (in Honan province, north of the Yellow River). The people then sang in his praise,

> Yeh has a good administrator,
> His name is Shih.
> He opened the river Chang to irrigate Yeh,
> What used to be old alkali land now can grow rice and *kaoliang*.[1]

This whole early period of the development of public works for water-control culminated in the achievement of the Chengkuo Canal, which bridged the transition from the period of feudal warfare to that of a united semi-feudal empire. Reserving the consideration of the details of its construction to a later chapter, suffice it to say here that it was opened in 246 B.C., the year which witnessed the crowning of the future First Emperor as the King of Ch'in.

Converting the Chin valley into fertile and flourishing farm land, the Chengkuo Canal laid the material basis for the prosperity and power of Ch'in and made central Shensi the first Key Economic Area in China. The control of this area gave Ch'in a powerful weapon for the subjugation of the rest of the feudal states.[2]

The resulting unification of China by Ch'in opened a new epoch in Chinese history, and ushered in a period in which the development of public works for water-control played an outstanding role in the political unity and division of China for twenty centuries, and greatly influenced the southward shifting of the centre of gravity of economic life from the Huang Ho valley to the Yangtze basin.

[1] Pan Ku, op. cit., *chüan* 29, Book on Ditches and Furrows, p. 2.
[2] Cf. *infra*, pp. 75-78.

Water-control as a Public Function

The foregoing genetic survey of water-control in ancient China reveals the significant fact that large-scale enterprises of this kind were from their very inception a function of the state. Large water-control enterprises were almost exclusively public works. This fact has been so much an integral part of the water-control system of China that it has usually been taken for granted. However, the historical and socio-economic implications of water-control development can never be fully realized unless the nature of its relation to the state and the manner in which it functioned are clearly stated. Karl Marx answered the question in regard to the Orient in general, in saying that:

> This prime necessity of an economical and common use of water, which, in the Occident, drove private enterprise to voluntary association, as in Flanders and Italy, necessitated, in the Orient where civilization was too low and the territorial extent too vast to call into life voluntary association, the interference of the centralizing power of government. Hence an economic function devolved upon all Asiatic governments; the function of providing public works.[1]

When it is realized that the development of voluntary associations for irrigation in Flanders and Italy occurred when merchant capital had already attained a high level, the meaning of Marx's comment that the civilization of the Orient was "too low" for a similar development is clear. In China, the growth of merchant capital has repeatedly been nipped in the bud by the ruling landlord bureaucracy, who promptly declared every important profitable enterprise a state monopoly

[1] Karl Marx, "The British Rule in India," *New York Daily Tribune*, 25th June 1853, p. 5.

and absorbed the embryonic merchant class into their ranks if it were felt it would be unsafe to disregard it.

Thus no sufficient capital was left among the people, or the non-bureaucratic section of society, and the foundation for any large scale voluntary association involving a large outlay of capital was destroyed. Hence the largest unit of co-operation was the village, and almost any enterprise beyond the capacity of the peasants of a single village required the interference of the magistrate, upon whom devolved the duty of mobilizing forced labour, supervising the construction of local works, and regulating the use of water by rival villages.[1]

WATER-CONTROL DUTIES OF THE OFFICIALS

The importance of the water-control duties of the local official or magistrate can be seen from the remark of one of the outstanding water-control authorities under the Ch'ing Dynasty. Says K'ang Chi-t'ien,

> As to the construction and the repair of canals in the interior, for the purpose of letting various people share in the benefits, the expenses are heavy and maintenance difficult. Although the officials could not do the work without burdening the people, yet, if it is left to the people, not only could it not be undertaken by one man or a single family, but it could easily give rise to quarrels; and furthermore [their meagre] resources would usually face quick exhaustion.[2]

Ku Shih-lien, a scholar and official who lived at the beginning of the Ch'ing dynasty, wrote an essay on the duties of the local magistrate in regard to irrigation and drainage. In setting forth what a good local

[1] It is significant that in Shansi province, where commercial capital was considerably developed during the Yüan, Ming and Ch'ing dynasties, privately constructed water-control works abound. Cf. *supra*, p. 44.

[2] K'ang Chi-t'ien, op. cit., *chüan* 4, p. 12.

official's water-control duties should be, the essay reveals the close relationship between the productive capacity of a certain region and the diligence of the local officials. Part of the essay reads :

> The magistrate is an official close to the people, and flood and drought should be of as much concern to him as pain or sickness of his person. Rivers, canals and dikes within a hundred *li* can be reached easily. During the intervening period between the busy seasons on the farm, the magistrate should visit the countryside, inquiring about the sufferings of the peasants and asking about their diligence or laziness. He should survey the topography of the region, ask about conditions of drainage, and investigate sluices and locks. He should find out what beneficial establishment should be promoted, what harmful things should be removed, where good crops have been harvested and where land has been deserted and reduced to waste : all these affect the conditions of the public treasury and the welfare of the people and must be carefully considered by the magistrate. . . .[1]

The tendency of local magistrates to evade this heavy responsibility can be seen from a letter in which the famous Sung official and scholar, Fan Chung-yen (989-1052), discussed " water-benefits " with the chief ministers. He blamed the local magistrates for shifting their responsibilities and relying too much on the initiative of the court, and expressly stated that " the duty of looking after the matter of ditches and sluices rests with the officials of the districts and prefectures." [2] However, when the works involved several districts, or when the expenses were such that higher officials felt an urge to dip their fingers in the pie, they were supervised by governors or even special imperial commissions.

[1] Ku Shih-lien, " Five Essays on Water-benefits," p. 5, published as part of *Ti-Hsiang-chai Ts'ung-shu.*

[2] *The Gazetteer of Su-Chou* [Soochow, Kiangsu], (1881 Edition), *chüan* 9, p. 4.

Frequently, imperial edicts were issued either urging a general speeding-up of water-control activities or ordering construction of specific works. For instance, the construction of transport canals, involving tens of thousands of labourers, working under the corvee-system, would usually be done under an imperial edict and supervised by an imperial commissioner. In the case of the embankments on the Yellow River, especially during the Ming and Ch'ing dynasties, the works were in charge of a special official ranking high in the bureaucratic hierarchy.

The seriousness with which officials considered their weighty and difficult task is revealed by the remark made by a high official of the Ming dynasty that "building embankments on the Yellow River is like constructing defenses on the frontier, and to keep watch on the dike is like maintaining vigilance on the frontier."[1] The vastness of the territory mentioned by Marx naturally accentuated the difficulty of the task. It should be obvious why works of such magnitude and gravity are beyond the capacity of the peasant or private merchants, and cannot be accomplished except by use of the centralized resources and authority of the state.

This dependence of the mass of the people upon the functioning of the bureaucracy was formulated by Max Weber as follows:

> . . . in the cultural evolution of Egypt, Western Asia, India, and China the question of irrigation was crucial. The water question conditioned the existence of the bureaucracy, the compulsory service of the dependent classes, and the dependence of the subject classes upon the functioning of the bureaucracy of the King.[2]

[1] *The Gazetteer of Honan*, *chüan* 4, p. 19.

[2] Max Weber, *General Economic History*, London: George Allen & Unwin, Ltd., 1927, p. 321.

Weber fails to bring out the important point that the functioning of the bureaucracy was conditioned by the political objectives of the ruling group rather than a general feeling of obligation to the mass of the people.

CHAPTER V

THE LOESS REGION AND CENTRAL HUANG HO BASIN AS A KEY ECONOMIC AREA

As China emerged from the classical feudal period in the third century B.C., unity was achieved for the first time when the First Emperor of the Ch'in dynasty conquered the various feudal states and brought the whole of China under one rule.[1] The state of Ch'in grew strong and powerful in the valleys of the Chin and Wei Rivers, the territory of the present Shensi province in the heart of the loess region in North China. Ssu-ma Ch'ien significantly attributes the secret of its success to the development of irrigation in the Ching River valley, particularly the cutting of the Chengkuo Canal which was opened in 246 B.C.,[2] twenty-six years before the First Emperor unified the country.

CH'IN AND THE CHENGKUO CANAL

The story of the Chengkuo Canal is a paradox, but paradoxes not infrequently occur in times of great change when confusion reigns in the minds of rulers. In the "Book on Rivers and Canals" in the *Historical Records*,[3] the story is told that:

In order to prevent the eastward expansion of Ch'in by tiring it out with other activities, Han sent the water

[1] Ku Chieh-kang correctly points out that the supposed unity of China in the earlier dynasties was historically untrue. See *Ku Shih Pien, Discussions in Ancient Chinese History*, Peiping, 1930, vol. ii, pp. 1-16.

[2] Ssu-ma Ch'ien, op. cit., *chüan* 15, Chronological Table of Six States, p. 425. The Table says, "First year of the First Emperor." What it is referring to is the first year when the man named Cheng, who later became the first Emperor, ascended the throne and became the King of Ch'in. This is indicated by the numbering of the years as Yi-Mao which the historian also employed. The year Yi-Mao is 246 B.C., not 221 B.C., when Cheng became Emperor.

[3] Ssu-ma Ch'ien, op. cit., *chüan* 29, p. 3.

(irrigation) engineer, Cheng Kuo, to Ch'in to persuade the [King of] Ch'in to open the Ching River and construct a canal from Chungshan [1] west to Huk'o and east of Peishan, carrying the water to the Lo River.[2] The proposed canal would be 300 *li* long and was to be used for irrigation. Before the construction work was finished, however, Ch'in authorities became aware of the trick and intended to kill Cheng Kuo, who then said to them, "Although the project was intended as a trick, yet the finished canal would be of great benefit to Ch'in."[3] The work was then ordered to be continued. After completion, it irrigated over 40,000 *ch'ing* of alkali land with water laden with rich silt. The productivity [of these fields] rose to one *chung* for each *mu*. Thus Kuanchung [present Shensi province] became a fertile country without bad years. Ch'in, then, grew rich and strong and finally conquered all other feudal states. The canal was [named after the engineer and] called Chengkuo Canal.[4]

Looking at the story in its historical perspective, it seems clear that the curious circumstances about the beginning of the enterprise are not accidental. The very fact that the authorities of Han (a state during the period of Warring States, not the Han dynasty of a later period) adopted such a scheme to fool Ch'in, indicates that the construction of large-scale public works for irrigation was not an established practice at that time, and that its pros and cons were not yet clear to the rulers of the various feudal states. Ch'in, which was considered an upstart and was known for its interest in innovations which had not yet been seasoned by experience, was thought of by the politicians in

[1] Present Chingyang district in Shensi province.

[2] The present Ch'i and Chü Rivers in Feng-Hsiang district in Shensi province.

[3] The *Earlier Han History* version has the following sentence in addition to the quotation in Ssu-ma Ch'ien's *Historical Records*: "'. . . I have, by this trick, prolonged the life of the State of Han for a few years but I have also done great service to Ch'in, the benefit of which Ch'in could enjoy for ten thousand years to come.' The Ch'in authorities agreed with him. . . ." Pan Ku, op. cit., *chüan* 29, "Book on Ditches and Furrows," p. 3.

[4] Ssu-ma Ch'ien, op. cit., Book 29, p. 1.

Han as an easy victim for a grandiose scheme of canal construction for irrigation.

Ch'in fell in with the scheme, but not exactly to its disadvantage. Tens and hundreds of thousands of peasant labourers of military age were employed for this work and were therefore prevented from swelling the rank of the already strong Ch'in army. Ambitions of conquest on the part of Ch'in might therefore have been checked somewhat during the years of public works construction ; but once this gigantic irrigation system was completed, running through the territory of Ch'in (the present Shensi province) as the blood system runs through a human body, it filled the granaries of Ch'in to overflowing. Ssu-ma Ch'ien understood the fundamental importance of the increase of productivity when he directly related the success of Ch'in to this project of irrigation development. Han paid with its existence as a state for ignoring this fundamental fact of history.

The Chengkuo Canal laid the foundation for the irrigation system in central Shensi for many centuries to come. It made central Shensi the Key Economic Area in China, the control of which gave Ch'in a weapon which in the end proved powerful enough to subjugate the rest of the country. But the control of the Key Economic Area merely means the gaining of a material advantage, a condition for success ; it neither guarantees success, nor insures against subsequent defeat after a temporary victory. Political and socio-economic conditions may give rise to a rebellious movement capable of driving the ruling power out of the Key Economic Area and taking its place.

This was exactly what happened to Ch'in. Legalizing private land ownership for the first time in Chinese history, and organizing the country on the basis of *chün* and *hsien* (districts and prefectures),[1] Ch'in released the rising private landowning class from the yoke of classical feudalism, thus hastening the process of class differentiation among the peasantry and increasing the pressure on the toiling masses who actually tilled the land. The result was that by the time of the reign of the First Emperor so many hundreds of thousands of peasants were already deprived of a living on the land, and such an army of surplus labourers had been created that the ruling group was over-tempted to inaugurate gigantic public works, such as the A-Fang Palace and the Great Wall, and to engage in foreign conquests as illustrated by the war of expansion against the Tartars. The consequence is well known. A powerful popular uprising, growing in strength concurrently with the spread of an anti-Ch'in movement organized by the nobles and satellites of fallen feudal states, overthrew the rule of Ch'in and, after years of struggle between different factions of the anti-Ch'in combination, the Han dynasty (206 B.C. to A.D. 221) emerged as the new ruler of a united China, with Kuanchung, the Key Economic Area taken from Ch'in, as the geographical and economic base of its power.

Kuanchung and the Rise of Han

After the fall of the Ch'in dynasty, Liu Pang, the founder of the Han dynasty, owed much of his success

[1] The historic move was made in 350 B.C., the twelfth year of the reign of Hsiao Kung, by which the Well-land system was officially pronounced dead and transfer of land by sale and purchase made legal, and the country of Ch'in was administratively divided into forty-one *hsien* or districts, each one ruled by a magistrate appointed by the king. Two years later, 348 B.C., the Fu, a system of land tax, was inaugurated. Ssu-ma Ch'ien, op. cit., *chüan* 5, "Biography of Kings and Emperors of Ch'in," pp. 22-23; and *chüan* 68, "Biography of Shang Yang," pp. 4-5. The abolition of the Well-land system was not expressly mentioned in Ssu-ma Ch'ien but was so stated in Pan Ku, op. cit., *chüan* 24, "Book on Food and Commodities," p. 7.

in finally defeating his powerful rival, Hsiang Yü of Ch'u, to his control of Kuanchung. The administration of this key area he had wisely entrusted to his chief political lieutenant, Hsiao Ho, who provided the Han army with a steady flow of supplies during the most critical period of the campaign. Just how important his contemporaries considered the work of Hsiao Ho is shown by the fact that, when the merit of the achievements of the various chief supporters of Liu Pang was discussed at court after the founding of the dynasty, that of Hsiao Ho was ranked first.

The reason for this award was expressly stated by its sponsor, Ao Ch'ien-ch'iu, an outstanding official of his time. He said, "When the armies of Han and Ch'u held their ground against each other at Jungyang (the present Jungyang district in Honan province) for several years,[1] the army [of Han] had no ready supply of food. Hsiao Ho despatched grain from Kuanchung by water transport and supplied food [for the army], thereby preventing a shortage. Your majesty lost Shantung several times, but Hsiao Ho continuously held Kuanchung at your disposal. This is an achievement which will endure for ten thousand generations."[2]

The key importance of Kuanchung in peace as well as in war was also recognized by Chang Liang, the brilliant political tactician in Liu Pang's camp. He characterized Kuanchung as a "fertile country,"[3] and states that "it commands the abundance of Shu and

[1] Three years. See Ssu-ma Ch'ien, op. cit., *chüan* 130, "Autobiography of Ssu-ma Ch'ien," p. 20.

[2] The Chinese phrase implies an infinite time. Pan Ku, *Earlier Han History*, *chüan* 39, "The Biography of Hsiao Ho," p. 4.

[3] The T'ang commentator, Yen Shih-ku, said that "the land is called fertile because it enjoys the benefit of irrigation." See Pan Ku, op. cit., *chüan* 40, "The Biography of Chang Liang," p. 8.

Pa (the present Szechwan)[1] in the south and the nomadic herds of the Hu in the north; and with [natural] barriers on three sides, it can be easily defended. With one side opening to the east, [it is an excellent base] for the subjugation of the feudal lords. After the feudal lords are subjugated, the Yellow and the Wei Rivers could be used to transport grain tribute westward to supply the capital. If the lords rebel, [army and supplies] could be despatched downstream. This is why Kuanchung can be designated as a metal city of a thousand *li* and a country that is a heavenly storehouse."[2] On the basis of this analysis, Chang Liang recommended Kuanchung as the seat of the capital.

Ssu-ma Ch'ien gives us a rather exact idea of the economic predominance of Kuanchung during the Han dynasty when he says that "Kuanchung occupies one-third of the territory under heaven[3] with a population three-tenths of the total; but its wealth constitutes six-tenths [of all the wealth under heaven]."[4]

WATER-TRANSPORT AND IRRIGATION DURING WU TI'S REIGN

The ruling power, however, was never disposed to rely for its support upon the supplies of the Key Economic Area alone. Whenever the country remained united and peaceful for some time, the appetite of the ruling group developed, its demand for goods to satisfy its mounting wants increased, its desire for

[1] Developed by a big irrigation project constructed by Li Pin during the Ch'in dynasty. Cf. *infra*, pp. 96-97.

[2] Pan Ku, op. cit., *chüan* 40, "The Biography of Chang Liang," p. 8.

[3] Meaning China.

[4] Ssu-ma Ch'ien, op. cit., *chüan* 129, "Biographies of Merchants and Industrialists," p. 7.

conquest of outlying territories also increased, and the necessity for new sources of revenue became more urgent.

During the Earlier Han dynasty the supplementary economic area which supplied the capital with grain tribute was the lower Yellow River valley,[1] which produced a tribute of several hundred thousand *tan* at the beginning of the dynasty.[2] This had increased to the alarming amount of 6,000,000 *tan* by 110 B.C.[3] There is small wonder, then, that the question of agricultural productivity in Kuanchung and the development of water transportation on the Yellow and Wei Rivers loomed large in the eyes of the authorities, especially during the reign of Wu Ti (140-87 B.C.).

Between 134 B.C. and 131 B.C., the Minister of Agriculture, Cheng Tang-shih, proposed that a canal of over 300 *li* (100 miles) should be cut to connect Ch'angan, the capital, with the Yellow River, so as to shorten by two-thirds the distance to be covered in transporting the grain tribute and reduce the time required for the journey by one-half.[4]

In making the recommendation the Minister was careful to point out the double purpose of the project, which is a common feature of a large number of water-control public works in China. He says that, apart from transportation, the canal " can also be used to irrigate some 40,000 *ch'ing* and more of the people's

[1] Called Kuantung, comprising southern Shansi, northern Honan and western Shantung.

[2] Pan Ku, op. cit., *chüan* 24, " Book on Food and Commodities," p. 8.

[3] Ssu-ma Ch'ien, op. cit., *chüan* 30, " Book on Control of Business," p. 18.

[4] Ssu-ma Ch'ien, op. cit., *chüan* 29, " Book on Rivers and Canals," p. 4. The Minister's words are : " Formerly, when Kuantung grain tribute was transported northward through the Wei River, it took six months to cover the 900 *li* and more, many difficulties being encountered on the way. If we can cut a canal to connect the Wei River with the Yellow River, the distance would be 300 *li* and three months would be long enough for the journey."

land. Thus," he concludes, "it will save [the trouble of] transportation and spare the soldiers, and obtain [additional] grain by enriching the land in Kuanchung."[1] Wu Ti approved this project and instructed Hsü Po-piao, water engineer from Ch'i, to take charge of the construction. History records that "several tens of thousands of soldiers"[2] were drafted to do the work. It was completed in three years.

About the same time, Fan Ssu, the governor of Hotung (east of the river, in modern Shansi), proposed that a canal should be constructed to bring the waters of the Fen River to irrigate P'ishih (the modern Hoching district, Shansi province) and Fenyin (the modern Jungho district, in Shansi province), and to bring the water of the Yellow River to irrigate Fenying and P'upan (the modern Yungchi district in Shansi province).

In support of his proposal, the governor said, "Over a million *tan* of grain tribute is transported from Shantung westward annually. Because of the difficulty of passing Tichu,[3] much loss is incurred in addition to the expensive cost of transportation. Hence the proposal to cut an irrigation canal that will irrigate approximately 5,000 *ch'ing* of land, adjacent to the river, which was formerly used by people to grow forage and for pasture. If irrigated and planted with crops it would produce over 2,000,000 *tan* of grain. The grain could be transported through the Wei River northward as if it were produced in Kuanchung itself. It would, then, not be necessary to transport the grain from east of Tichu. . . ."[4]

[1] Ssu-ma Ch'ien, op. cit., *chüan* 29, p. 81. [2] *Ibid.*

[3] A stone barrier in the middle of the Yellow River at the southern border of Shansi.

[4] Ssu-ma Ch'ien, op. cit., *chüan* 29, p. 81.

The emperor Wu Ti approved the project and ordered several tens of thousands of soldiers to work on the canal. A few years later, a change of the course of the river spoiled the canal and finally all the reclaimed land was again laid waste.

It is rather significant that the two instances of water-control public work construction credited to the early years of Wu Ti's reign were both undertaken primarily to facilitate transportation. In a state whose aim is the appropriation and concentration of grain tribute, this is natural. It was only when the government was hard pressed for more and more grain, and when over-appropriation and under-production were driving the peasants to unrest and rebellion, that necessity forced it to look toward the development of production for a solution of its problems, and to engage in irrigation works as an object of its primary concern.

Wu Ti's Edict on Irrigation

In 111 B.C., Wu Ti expressed his belief in the importance of irrigation by issuing the following edict:

> Agriculture is the basic [occupation] of the world. Springs and rivers make possible the cultivation of the five grains. . . . There are numerous mountains and rivers in the domain, with whose use the ordinary people are not acquainted. Hence [the government] must cut canals and ditches, drain the rivers and build dikes and water tanks to prevent drought.[1]

This edict referred to the petition of Erh Kuan, the Left Internal Adviser, who had proposed to cut six canals supplementary to the Chengkuo Canal for the irrigation of land adjacent to the main canal, but too

[1] Pan Ku, op. cit., *chüan* 29, "Book on Ditches and Furrows," p. 7.

high to be irrigated by it.[1] In this decree, Wu Ti laid bare the basic problem of agricultural production in China, namely, the proper use and regulation of water, and acknowledged the major economic function of the state, in regard to production, to be the construction and maintenance of water-control public works. A proper understanding of these two problems has always been a prerequisite of constructive statesmanship in China.

Flood-control and Key Economic Area

When the six supplementary canals were built, the Yellow River had been flooding Honan and Shantung for some time and the dike, which had burst at Kutze, had not yet been repaired. The fact that Wu Ti, instead of devoting his attention first to the apparently more important task of repairing the Yellow River dikes, should allow the cutting of the six supplementary canals to take precedence, enraged a historical commentator of the Ch'ing dynasty. K'ang Chi-t'ien criticized Wu Ti for neglecting the welfare of the people of five or ten *chün* (prefectures) by leaving the dike of the Yellow River unrepaired while devoting his attention to an enterprise that "only benefited a corner of the realm." [2]

K'ang's criticism, while completely justified on humanitarian grounds, betrays his ignorance of the pivotal importance of the Key Economic Area in the economic policy of semi-feudal China. It is true that the six supplementary canals benefited merely a

[1] The petition was granted and the canals were cut.

[2] K'ang Chi-t'ien, *Notes on Rivers and Canals* (*Ho-chu Chi-wen*), 1804, *chüan* 3, p. 8.

" corner " of the realm, but it happened to be a very important corner, important enough to be regarded as the Key Economic Area. The five or ten *chün* destroyed by flood perhaps had a much larger total population and greater total yield of grain, but they happened to be far away from the capital. Wu Ti's estimate of the relative importance of the two areas and the two water-control public works, when he approved the plan to construct the six supplementary canals, indicates how well he understood the significance of the Key Economic Area.

In order to strengthen the economic position of Kuanchung, Wu Ti also built the Paohsieh Road which connected the Pao and Hsieh Rivers at a distance of about 100 *li* or about 33 miles. The original intention was to transport tribute grain from Nanyang in the Han River basin through the Mien River to the Pao, then overland by the Paohsieh Road to the Hsieh River, and then through the Wei River to the capital. The road was built, but the presence of a large number of boulders in the rivers prevented their use for navigation.

On top of this failure to tap the resources of the Han basin, the Emperor met with another disappointment in the construction of the Lungku Canal, which was intended to harness the Lo River for irrigation. Ten thousand soldiers were engaged in the enterprise, but " after over ten years of construction, although the canal was cut through, no benefits could yet be gained from it."[1] Disappointed at the failure of these two enterprises, Wu Ti turned his attention to the dikes of the Yellow River at Kutze, and they were finally repaired under his personal supervision.

[1] Pan Ku, op. cit., *chüan* 29, " Book on Ditches and Furrows," p. 5.

This was twenty years after the dike had burst and the successful repair of the breach was deservedly lauded by later historians as a great achievement.

The personal role which the Emperor played in this enterprise, and the ceremonious acclaim that welcomed its success, duly impressed on the bureaucracy of the nation the importance of the construction of water-control works as a public function.

IRRIGATION WORKS IN THE NORTH-WEST AND THE HWAI VALLEY

"Henceforth," to quote the historian Pan Ku, "those in power all rivalled each other in recommending 'water-benefits.' In Shuofang, Hsiho, Hohsi, and Chiuch'üan prefectures,[1] the water from the Yellow River and its tributaries was used for irrigation. In Kuanchung, the Lintzu, Chengkuo, and Wei [2] canals took water from various rivers and streams. In Junan and Chiuchiang prefectures, water was led from the Hwai River, while in Tunghai [3] it was taken from the Chuting Lake. The land under T'aishan took water from the Wen River. Canals for irrigation were cut in each of these cases, and each canal irrigated over 10,000 *ch'ing* of land. Small ditches and petty dams to catch torrents from mountain slopes were too numerous to be mentioned."

The range of territory covered by this summary of irrigation projects is very interesting. It shows that

[1] All situated along the northern bend of the Yellow River, in Kansu, Ninghsia and Suiyüan provinces.
[2] Not the same Wei as the Wei River.
[3] Yen Chou, Shantung province.

to the north-west, agricultural reclamation work was carried on in what are now the provinces of Kansu, Ninghsia and Suiyüan,[1] while in the south-east, the Wei and Hwai Valleys were developed.[2] This undoubtedly laid the foundation for the further development of these regions toward the end of the Han dynasty.

The Po Canal and other Works

The crowning success of the period, however, was the Po Canal, in the heart of the Key Economic Area of Kuanchung. It was a reconstruction of the Chengkuo Canal which had become silted up and practically useless. The Po Canal was cut in 95 B.C., 160 years after the opening of the Chengkuo Canal, and sixteen years after the construction of the six supplementary canals at the proposals of the official Po Kung.

It tapped the Ching River at Kukou (the modern Chingyang district), a place much further upstream than the beginning of the Chengkuo Canal. Following a higher grading than its predecessor, it flowed to the Wei River at Yüeh Yang.[3] The canal was about 200 *li* long, irrigated over 4,500 *ch'ing* of land, and was named the Po Canal after its sponsor, Po Kung. The song, which the people were said to have sung in happiness and gratitude over the benefits they derived from the canal, has since become a classic, frequently quoted

[1] The site of the Han Canal can still be traced. See the *Gazetteer of Shuo-Fang*, *chüan* 2, p. 7, and Map 35 in *The Atlas by Provinces, Series A* by Ou-yang Yin, Wuchang, 1933, showing irrigation canals in Ninghsia province.

[2] These valleys were considerably developed during the preceding epoch, in the periods of the Spring and Autumn and Warring States.

[3] See K'ang Chi-tien, op. cit., *chüan* 3, p. 12.

and referred to by later writers as evidence of the beneficial effects of irrigation construction :

> Where is the land ?
> At the mouth of Ch'ih Yang Gorge.
> Chengkuo set the example,
> Po Canal followed suit.
> There is cloud when you raise the pick,
> And you get rain when the dike is cut.
> A *tan* of Ching water contains much silt ;
> It irrigates and it fertilizes ;
> It makes my crops grow ;
> And feeds millions in the country's capital ! [1]

Most of the above material concerning the water-control developments of the Emperor Wu Ti is taken from the " Book on Ditches and Furrows " in Pan Ku's *Earlier Han History*. Ssu-ma Ch'ien's *Historical Records* contain similar accounts, though not as complete. In the " Book on Finance," of his *Historical Records*, Ssu-ma Ch'ien recorded the cutting of the Chih Canal which connected Ch'angan, the capital, with Huayin. He also refers to the cutting of a large canal in Shuofang which employed " tens of thousands of workers." Both enterprises took two or three years, and, " before the works were completed, tens of thousands of gold were spent." [2]

Ssu-ma Ch'ien considers that it was this and other similar expenses that caused the financial difficulties in Wu Ti's reign, and brought about the issue of " deer skin " money. It is not surprising, then, that after the reign of Wu Ti, no major water-control public works were constructed during the Han dynasty, either directly by the central government or with its express approval and under its supervision. The inability of

[1] Pan Ku, op. cit., *chüan* 29, " Book on Ditches and Furrows," p. 8.
[2] Ssu-ma Ch'ien, op. cit., *chüan* 30, " Book on Finance," p. 6.

the central government to finance major public works for water-control after Wu Ti's reign resulted in the neglect of irrigation development in Kuanchung and the consequent economic decline of that Key Area.

WATER-CONTROL WORKS IN HWAI AND HAN VALLEYS AFTER WU TI

Some notable local enterprises, however, were developed on the initiative of energetic local administrators in the post Wu Ti period. It is highly significant that most of these enterprises were concentrated along the northern tributaries of the Han River and the upper course of the Hwai River in what is now Honan province, indicating a tendency which forewarned the rising of a rival economic area that would challenge the supremacy of the declining Kuanchung. Between 38 and 34 B.C., when Shao Hsin-chen was governor of Nanyang (southern Honan), he ordered the building of the Kanlu Reservoir by piling up stones to dam up the water of the Yü River, a large northern tributary of the Han River. Six stone sluices were constructed at the side of the dam to regulate the water supply. The area thus irrigated from the reservoir totalled 20,000 *ch'ing*.[1]

This work is known in connection with the first system of regulation of water-rights recorded in Chinese history and its author, Shao Hsin-chen, is regarded as the father of water-right regulation in China. Besides this work of historical fame, many other reservoirs and dams, scattered over the whole

[1] Pan Ku, op. cit., *chüan* 89, "Biographies of Good Administrators," pp. 14-15.

of southern Honan, were constructed at about the same period.[1] The *Earlier Han History* also refers to a reservoir called the Hungch'itapei, which enriched the prefecture of Junan in south-eastern Honan and western Anhwei.[2]

FLOOD-CONTROL AND DEVELOPMENT OF HONEI

In the same period when irrigation developments in Kuanchung were neglected by the central government, and enterprising local administrators were building irrigation works along the northern tributaries of the Han River and the upper course of the Hwai River, history also records extensive developments of flood-control works in the lower Yellow River valley. A large part of the "Book on Ditches and Furrows" in Pan Ku's *Earlier Han History* is devoted to recording the elaborate discussion at Court of the problem of Yellow River flood-control.[3] From some of the discussions, especially the famous memorial to the throne submitted by Chia Jang, one can get a glimpse of water-control development and agricultural prosperity at that period in what is now Honan province. Chia wrote his memorial at the royal command of Ai Ti (6-1 B.C.). This is how he described the complicated

[1] K'ang Chi-tien, op. cit., *chüan*, 3, p. 16.

[2] The reservoir was out of repair and was causing flood in the region when it was destroyed by order of Chen Ti (32-7 B.C.) at the recommendation of the minister Chai Fang-chin. After Chai's fall from power, villagers who were formerly benefited by the reservoir, accused Chai of wantonly destroying it for spite, because his request for a grant of rich land below the reservoir had been refused. During the reign of Wang Mang (A.D. 9-23) the country suffered much from drought. The sentiment against Chai was still strong, and folk-songs attacking him were sung by the peasants.

[3] Discussions on the technical phases of control, in which the records abound, are beyond the scope of this book.

system of flood-control in northern Honan at the time of his petition :

. . . Now, narrow embankments stand at a distance of several hundred steps from the water, and even wide ones only several *li* from it. Near the south of Liyang (the Hsün district in modern Honan) the old Tach'ing Dike stretched north-westward from the west bank of the Yellow River to the southern foot of the Western Mountain, then it turned eastward to meet the Eastern Mountain. People built their cottages on the eastern side of the dike. After living there a little over ten years they also built a dike from the Eastern Mountain southward, to connect it up with the great dike. In the prefecture of Neihuang (the modern Neihuang district in northern Honan) a swamp with a circumference of several tens of *li* drained by building a dike around it. The governor of the region gave the land within the dike to the people after they had lived there for over ten years. Now people build cottages in it. This is what I have witnessed myself. In the prefectures of Tungchün and Paima (southern Hopei and south-western Shantung) there are several rings of embankments, and the people live in between them. From the north of Liyang to the border of Wei (the former State of Wei) the old great embankment lies several tens of *li* from the river ; and inside the embankment there were also several rings of dikes which were built in earlier generations. Thus, when the Yellow River flowed from Honei northward to Liyang, there was a stone embankment to force it eastward. When it reached Tungchün and P'ingkang there was a stone embankment to force it north-east. When it came to Liyang and Kuan it was forced to flow eastward by another stone embankment. As it reached Tungchün and Chinpei another embankment forced it north-west, and in Weichün and Chaoyang one more stone embankment turned the river north-east again. Thus, in a distance of over a hundred *li* (about 33 miles), the river was turned twice westward and three times eastward. . . .[1]

Chia Jang's purpose in giving this account of contemporary conditions in northern Honan and southern Hopei was to show the over-embankment of the river as an aggravation of the difficulties of flood-control.

[1] Pan Ku, op. cit., *chüan* 29, "Book on Ditches and Furrows," pp. 13-14.

Nevertheless, the elaborate system of embankment, as described, must itself have been the result of an intensive effort to control floods. The magnitude of the works can be seen from the fact, as mentioned by Chia Jang in the same document, that "the ten *chüns* bordering the Yellow River spent millions annually for embankments."

When it is remembered that in the lower Yellow River valley the problem of water-control is not one of canalization, but one of flood-control and drainage, the significance of the elaborate construction of embankments as an indication of the increasing economic importance of the region can easily be realized. The neglect of canalization in Kuanchung and the elaborate construction of flood-control works on the Yellow River floods and the development of Honan indicate that the centre of gravity of economic policy had shifted, and that Honei, or what is now northern Honan, southern Hopei and western Shantung, had, by the end of the Earlier or Western Han dynasty (206 B.C. to A.D. 25), risen from the position of a supplementary base to replace Kuanchung as the main Key Economic Area.

Chia Jang's revelation of the chaotic manner in which the Yellow River embankments were constructed indicates that the effort of the local officials had not been co-ordinated by effective central direction. Thus it seems that, on the basis of the growing economic importance of the region, the political power of the local rulers had also increased, and had become more or less independent in fact, though not in name, of the central authority in Kuanchung.

HONEI AND THE FOUNDING OF THE LATER HAN DYNASTY

Liu Hsiu, the founder of the Later or Eastern Han dynasty, understood the meaning of the shift of the main Key Economic Area from Kuanchung to Honei. By a masterstroke of statesmanship he seized Honei at a critical stage of his campaign of conquest, and appointed one of his most reliable lieutenants, K'ou Hsün, to be its governor, at the same time authorizing him to exercise the power of a general.

In entrusting Honei to K'ou Hsün, Liu Hsiu said to him, "Honei is rich. It is the base from which I shall rise to conquer the country." Kao Tsu[1] appointed Hsiao Ho to guard Kuanchung: "I am entrusting Honei to your care. Guard it firmly and transport a sufficient supply of grain to my army. Lead and encourage your men and horses to block enemy forces from coming northward." [2]

When Liu Hsiu had united the country, after a short period of armed struggle against other contending groups who had rebelled against Wang Mang (the usurper whose reign (A.D. 9-23) marks the interval between the Earlier or Western Han and the Later or Eastern Han), he selected Loyang in the lower Yellow River valley, and not Ch'angan in the Wei River valley, as his capital.[3] This change is very significant. Although Kuanchung was economically still very important, the lower Yellow River valley and the Han and Hwai River valleys were destined to play an increasingly important role in Chinese history.

[1] Liu Pang, the founder of the Earlier Han dynasty.

[2] Fan Yeh, *Later Han History*, *chüan* 46, "Biography of K'ou Hsün," p. 19.

[3] Hence historians call his dynasty the Eastern Han in contrast to the Western Han, which had its capital at Ch'angan.

Two Reservoirs in the Hwai Valley

The Later Han dynasty which ruled China for 196 years, from 25 to 221, was not noted for its achievements in water-control. Fan Yeh's *Later Han History* contains no special book or chapter comparable to the chapters on the subject in Ssu-ma Ch'ien's *Historical Records* or Pan Ku's *Earlier Han History*. From the other parts of Fan Yeh's work and from other sources, a few items can be collected that will vaguely indicate the trend of development in this period.

There were attempts to continue the rehabilitation of the irrigation works of southern Honan and the Hwai River valley begun in the latter years of the Earlier Han dynasty. The *Biography of Wang Chin* records that, between A.D. 78-83 when Wang Ching was the governor of Luchiang (in Anhwei province), he rebuilt the Shih Reservoir, said to have been first constructed by Sun Shu-ao in the Spring and Autumn period (722-481 B.C.). As a result, the region enjoyed great abundance. Wang Ching also taught the peasants to use the plough for tillage.[1]

In A.D. 189 Ch'en Teng, the governor of Kwangling (southern Anhwei and northern Kiangsu province), constructed a reservoir which was controlled by a series of weirs running from Shou (in Anhwei province) eastward. Extending over a winding course of 90 *li*, about 20 miles, the reservoir collected the water of thirty-six streams from the mountain ranges which formed its catchment area in the north-west. It irrigated over 10,000 *ch'ing* of land.[2] The Hwai valley, however, did not attain a high level of

[1] Fan Yeh, op. cit., "Biography of Wang Ching," *chüan* 106, p. 8.
[2] K'ang Chi-t'ien, op. cit., *chüan* 4, pp. 19-23.

development, for reasons which will be discussed in the following chapter.[1]

KEY ECONOMIC AREA OF THE TWO HAN DYNASTIES

As far as the period of the two Han dynasties is concerned, the centre of agricultural productivity was still in the north of the Hwai valley. In an edict issued in A.D. 115, the Emperor, An Ti, incidentally gave a most comprehensive enumeration of the agriculturally most developed and richest sections of his empire. The edict ordered his ministers and governors to repair the old irrigation canals and water routes in Sanpu (central Shensi), Honei (northern Honan), Hotung (south-western Shansi), Shangtang (south-eastern Shansi), Chaokuo (western central Shansi and eastern central Hopei), and T'aiyüan (central Shansi).[2]

In other words, An Ti, in issuing this important edict, indicates that he was particularly concerned over productivity and transport facilities in the heart of the country. Although the Western Han emphasized Kuanchung, while the Eastern Han centred more of its attention on Honei, the whole region covering the valleys of the Ching, Wei and Fen, and the Honan-Hopei section of the Yellow River, constituted the Key Economic Area, the main basis of support and the seat of power during the whole period of the two Han dynasties, from 206 B.C. to A.D. 221.

[1] Cf. *infra*, p. 104.
[2] Fan Yeh, op. cit., *chüan* 5, "Biography of An Ti," p. 11.

CHAPTER VI

THE TRANSITION FROM THE HUANG HO BASIN TO THE YANGTZE VALLEY

THE fifty years of struggle between the Three Kingdoms (A.D. 221-265) which succeeded the Later Han Dynasty constituted the first period of sustained division in the semi-feudal epoch. Unlike other later periods of division which were complicated by the simultaneous occurrence of barbarian invasions, the period of the Three Kingdoms was a typical case of division generated by the internal forces of Chinese society.

The material and fundamental factor responsible for such division was the rise of rival economic areas, whose productivity and location enabled them to serve as bases for a sustained challenge of the authority of the overlord who commanded the central or main Key Economic Area. In this case it was the increasing maturity of Shu, or the Szechwan Red Earth Basin, and the adolescent exuberance of Wu, the lower Yangtze valley, that produced the balance of power politically represented by the Three Kingdoms.

WATER-CONTROL AND SZECHWAN

Shu, or Szechwan, began to acquire considerable importance in Chinese history after King Chao Hsiang (306-251 B.C.) of the state of Ch'in moved his people from Ch'in (Shensi) to colonize the territory.[1] The colonization of Shu was accompanied by the construction of a remarkable system of irrigation which established the basis for the prosperity of the region. The honour of being the father of the irrigation system in

[1] Pan Ku, op. cit., *chüan* 39, p. 2.

Szechwan fell to Li Ping, the Ch'in governor, whose work has been considered an immortal achievement by Chinese historians.

The central work in the system was the Tuchiang Dam in the Kuan district which divided the Min River into two main streams, each of which branched out into many minor canals.[1] The canals were primarily cut for purposes of transportation, but once cut, they were also extensively used for irrigation. "Small irrigation ditches amounted to hundreds of thousands,"[2] and the Ch'engtu plain thus was called "sea-on-land."[3]

The benefit which the people derived from the Tuchiang Dam and the canals was not limited to transportation and irrigation. A stone tablet of the Yüan dynasty definitely states that "water power stations for polishing rice, milling, and spinning and weaving, numbering tens of thousands, were established along the canals and operated for the four seasons of the year."[4]

It is no exaggeration to say that the Ch'engtu plain owes its fertility and economic self-sufficiency to this water-control system. Toward the end of the Earlier Han dynasty, Szechwan had also become sufficiently rich to enable Kung-sun Shu, military chieftain occupying the territory from A.D. 25 to 36, to defy, longer than any other rival group, Liu Hsiu's attempt to unify China. During the period of the Three Kingdoms it provided a base which allowed the Shu Han dynasty (221-265) to hold out against its enemies in North and Central China for nearly fifty years.

[1] See the map in the *Gazetteer of Szechwan Province*, *chüan* 1, pp. 11-13.
[2] K'ang Chi-t'ien, op. cit., *chüan* 2, pp. 32-33.
[3] *Ibid.*
[4] Yüan Chieh-hsi-ssu Shu Dam tablet inscriptions, copied in *Gazetteer of Szechwan Province*, *chüan* 13, p. 27.

Primitiveness of Yangtze Valley during the Han Dynasty

Ssu-ma Ch'ien characterized the lower Yangtze valley, known as Ch'u and Yüeh, as a

> large territory sparsely populated, where people eat rice and drink fish soup ; where land is tilled with fire and hoed with water ;[1] where people collect fruits and shellfish for food ; where people enjoy self-sufficiency without commerce. The place is fertile and suffers no famine and hunger. Hence the people are lazy and poor and do not bother to accumulate wealth. Hence, in the south of the Yangtze and the Hwai, there are neither hungry nor frozen people, nor a family which owns a thousand gold."[2]

This remarkable account, which gives practically all the essential facts for judging the economic development of a region, clearly describes the Yangtze valley during the Han dynasty as a country with a small and scattered population living under a commercial economy, with a primitive agriculture, no exchange of goods, and little sign of class differentiation.

Professor Liu Yi-cheng's researches reveal that, during the Eastern or Later Han, in the region around Huangchow and Tean in Hupei province, and even in some parts of Anhwei, there were a large number of "Southern barbarians," a very primitive people.[3] Even in the period of the Three Kingdoms many dis-

[1] Wu Ti mentioned this form of cultivation in Kiangnan (south of the Yangtze) in an edict. Ssu-ma Ch'ien, op. cit., *chüan* 30, p. 15. The reference is to a primitive form of agriculture, cultivating more or less virgin soil by burning down the overgrowth, flooding the land, and at about the same time, seeding rice. This sort of agriculture is still prevalent in some parts of Annam.

[2] Ssu-ma Ch'ien, *chüan* 102, "Biographies of Merchants and Industrialists," p. 12.

[3] Liu Yi-cheng, *History of Chinese Civilization*, Nanking, 1932, vol. i, p. 388.

tricts in Wulin (Changte in Hunan province) were still inhabited by "barbarians."[1] Sun Ch'üan, the founder of the Wu dynasty (229-265), one of the Three Kingdoms, had to force the barbarian inhabitants of his kingdom to serve in the army to make up the shortage of soldiers in the small population.[2]

There was, however, a sufficient number of settlers from the north who, carrying with them all the equipment of an advanced agricultural civilization, could take good advantage of the incomparable natural fertility of the Yangtze valley to establish an independent kingdom under the guiding hand of Sun Ch'üan, who forced the powers in the Yellow River valley and the Ch'engtu plain to treat the new State as their equal.

WATER-CONTROL AS A WEAPON IN THE THREE-CORNERED STRUGGLE

Undoubtedly, military campaigns were the most spectacular aspect of the three-cornered struggle under the Three Kingdoms. But behind the military fronts much effort was spent, especially by Shu Han and Wei, on developing agricultural productivity and water transport as a means of strengthening their military power.

Wu's chief problem was shortage of the man-power needed to develop the virgin fertility of natural resources still within easy reach. History records, however, that the King of Wu, in 226, decreed the establishment of military agricultural colonies, in order to extend the area under cultivation and make up a grain

[1] Ch'eng Shou, *History of the Three Kingdoms*, Part III; *History of Wu*, *chüan* 2, p. 9.

[2] Liu K'o-yü, *The Nine Encyclopedias of Source Material on the Institutional History of China*, *chüan* 19, p. 13.

shortage which was then a danger to the kingdom.[1] In 245 the King dispatched his general, Ch'en Hsün, with 30,000 soldiers and workers to cut the Chüyung Canal connecting Hsiaoch'i and Yünyang (the modern Tanyang district in Kiangsu), and made the Ch'ihshan Lake, in the vicinity of Tanyang, for the purpose of irrigating rice-fields.

In Shu Han the brilliant minister, Chu-ko Liang, fearing that the grain supply from his base in Szechwan might be cut off, established military agricultural colonies on the southern bank of the Wei valley in 234, when his attempt to conquer the Yellow River valley was checked by the Wei armies under Ssu-ma Yi.[2] Water-control public works are not expressly mentioned in the records, but taking into consideration the condition of agriculture in the regions involved, it is safe to assume that irrigation development was an indispensable factor in military agricultural colonization.

In the case of Wei, irrigation as well as military agricultural colonization was carried out most extensively, and the records are most explicit in regard to the relation between the two. Diligent attention was also paid to the extension and improvement of the routes of water transportation. Both kinds of enterprise were, for obvious reasons, centred in the Hwai River valley. The dams of the Shih Reservoir, the Ju Reservoir, the Seven Gates and the Wu Reservoir were repaired by Liu Fu at the order of T'sao T'sao, who appointed him governor of Yangchou.[3] In 204 T'sao

[1] Ch'en Shou, op. cit., *chüan* 2, "Biography of Sun Ch'üan," pp. 16-17.

[2] Ch'en Shou, op. cit., Part II, "History of Shu," *chüan* 5, "Biography of Chu Ke-Liang," p. 14.

[3] Fang Ch'iao, et al., *Tsin History*, *chüan* 26, "Book on Food and Commodities," pp. 6-7. According to K'ang Chi-t'ien, the Shih Reservoir was repaired in 208 by order of T'sao T'sao for purposes of military colonization. See *Notes on Rivers and Canals*, *chüan* 4, pp. 25-26.

T'sao deepened the Pien Canal, cut the Sui Canal to connect the Pien Canal with the Hwai valley, and dammed the Ch'i River (in northern Honan) to force it into the Peik'ai Canal, in order to bring the grain products of Shantung to the centre of his domain in Honan.[1] This was at a time when T'sao T'sao was engaged in a gigantic scheme of military colonization, storing up grain supplies and developing water-routes of communication.[2] The achievement of the year 204 in the Hwai and Yellow River valleys was followed in the next year by the cutting of the P'inglu Canal in western central Hopei province to lead the water of the Hut'o River into the Sha River, and the Ch'üanchow Canal in central Shansi to use the water of the Lu River for transportation. T'sao T'sao's victory over the Wuhuan barbarians in the north was largely due to the two canals which were dug as a part of the preparation for this conquest.[3] The Yang Canal in Honan province was cut in 219 to connect the Lo River with the Pien Canal for transporting tribute grain.[4] In 233 the Ch'engkuo Canal which ran from Ch'ents'ang (Paochi district in Shensi province) to Huaili (*Hsing-p'ing* district in Shensi province) was opened. About the same time the water of the Yen and Lo Rivers was utilized by the construction of the Linchin Reservoir to irrigate about 3,000 *ch'ing* of alkali land.[5]

As pointed out by K'ang Chi-t'ien,[6] these works must have been constructed on the initiative or under the direction of the great Wei general, Ssu-ma Yi,

[1] K'ang Chi-t'ien, op. cit., *chüan* 4, p. 24.

[2] See the comment by K'ang Chi-t'ien, op. cit., *chüan* 4, pp. 24-25.

[3] Both of the canals were of great service in the T'ang as well as Ming dynasties. K'ang Chi-t'ien, op. cit., *chüan* 4, p. 25.

[4] Fan Yeh, op. cit., *chüan* 65, "Biography of Chang Ch'un," pp. 2-3.

[5] Fang Chiao, et. al., op. cit., *chüan* 26, "Book on Food and Commodities," p. 8.

[6] *Notes on Rivers and Canals*, *chüan* 4, pp. 30-31.

who at that time was defending Wei against the invading Shu Han force led by Chu-ko Liang. Ssu-ma Yi was afraid of the military genius of his opponent, but he knew that the Shu Han army's chief weakness lay in its relatively inferior economic base. Therefore he refused to engage in direct combat with the enemy and tried to starve them out, conducting the war with spades, ploughs, irrigation canals, and reservoirs, rather than swords and spears. Chu-ko Liang, after failing to provoke his enemy to battle by all sorts of insults, had to use the same seemingly clumsy but highly effective weapons, and organized military agricultural colonies on the southern bank of the Wei River. Thus the war between Shu Han and Wei was turned into a contest of economic strength and endurance, and by evading the offensive strength of the Shu Han forces and putting his enemies at a great disadvantage, Ssu-ma Yi prepared the ground for the final conquest and the elimination of Shu Han.

Wei wisely adopted the tactics of facing one enemy at a time, and while bent on the conquest of Shu Han, peace and friendship were maintained with Wu. This political accord could be seen even in the field of irrigation development. History records that Chia Ku'ei, the Wei governor of Yüchow (in Honan province), secured the co-operation of Wu and utilized tools which had been prepared for military purposes to construct a dam on the Ju River (in southern Honan and northern Anhwei), as well as to build a reservoir called the Hsin Reservoir, and to cut a transportation canal of over 300 *li*. The canal, a monument of Wei-Wu friendship, was named the Chiahou Canal.[1]

[1] Fang Ch'iao, et. al., op. cit., *chüan* 26, "Book on Food and Commodities," p. 7.

However, the Wei-Wu partnership proved to be short lived. Even before Shu Han was completely subjugated, Wei made preparations to strengthen its economic base for the conquest of Wu. In 241, military agricultural colonies began to be established in the Hwai Valley; land under cultivation was greatly extended, canals were cut and transportation facilitated; and a large supply of grain was stored up.[1]

In 243, after the campaign against Wu was opened, the army of Wei, under the command of the King himself, defeated the Wu general, Chu-ko Ke, by burning his accumulated supplies. The King planned to strengthen his own economic reserve even further, and made preparation for future campaigns by a further extension of the area under cultivation and a greater accumulation of grain. He ordered his ablest general, Teng Ai, to the territory from the west of Ch'en (the modern Hwaiyang district in Honan) and Hsiang (the modern Hsiangch'en district in Honan) to Shouch'ung (the modern Shou district in Anhwei). General Teng knew that, although the land in that region was good, there was not enough water to take full advantage of the land, and he proposed to cut a canal both for the purpose of irrigation and for transportation. The plan was approved by the King. Camps of sixty soldiers were stationed 5 *li* apart from each other for over 400 *li*, from the bank of the Hwai River, starting at Chungli (the modern Fengyang district in Anhwei province on the southern bank of the Hwai River) to the Pi River (entering the Hwai River at Chenyüan Kuan). The soldiers were both to cultivate the land and to defend the territory; they were also to engage in the work of repairing and

[1] K'ang Chi-t'ien, op cit., *chüan* 4, pp. 28-29.

widening the Hwaiyang and Peichih Canals. Water from the Yellow River was thus led from the north down to the reservoirs along the Yin, Hwai and Tachih Rivers. Over 300 *li* of canals were dug both north and south of the Yin River, irrigating about 20,000 *ch'ing* of land.

Thus, " the south and the north of the Hwai were linked. From Shou Ch'ung to the capital in Hsüch'ang (the modern Hsüchou in Honan province) the noise of dogs and chickens in the fields of the government and of the soldiers, and on private farms, could be heard by each other, and the fields of one dove-tailed into those of the others. In times of military emergency in the south-east, when the great army sailed down to the Hwai and Yangtze River valleys, there would be food and other supplies stored up and there would be no fear of damage by flood." [1]

The system of irrigation thus established contributed greatly to the agricultural development of the Hwai valley. It represented the culmination of a whole period of energetic competitive increase in agricultural productivity for purposes of military struggle.[2] It added a rich productive area to the kingdom of Wei, and, thenceforth, Wei became so powerful that the balance of power between the Three Kingdoms was resolved, and China was again united under one rule.

Hwai Valley as the Historic Battlefield

But the great military importance of the Hwai valley, which gave impetus to these remarkable water-

[1] Fang Ch'iao, et. al., op. cit., *chüan* 26, " Book on Food and Commodities," p. 8.

[2] " From the Huang Ch'u period (221-226) to the Tsin dynasty, able ministers all considered the cutting of canals and storing of grain to be the means for military preparations." K'ang Chi-t'ien, op. cit., *chüan* 4, p. 27.

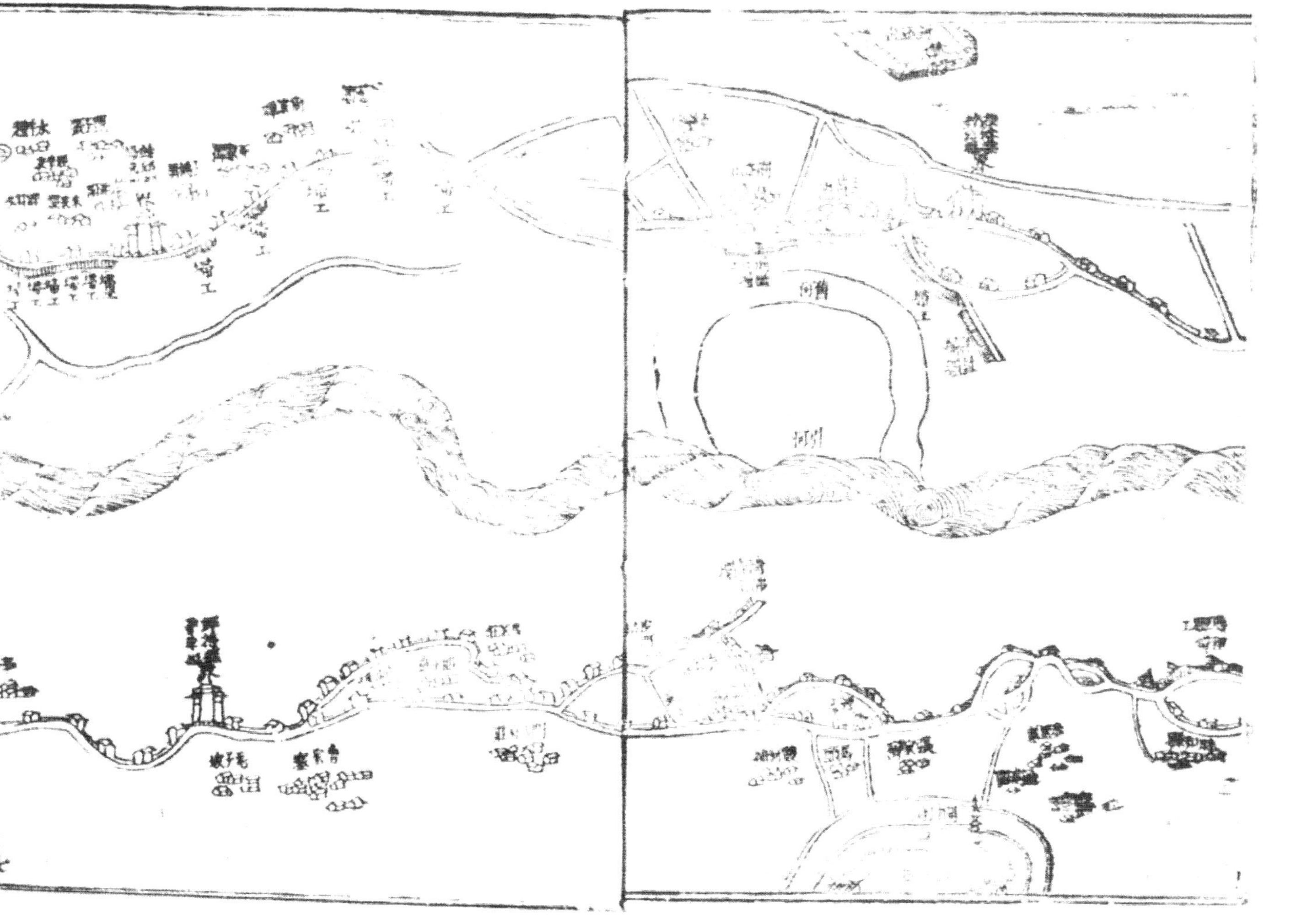

Embankment on the Yellow River, north of Kaifeng (Honan)—Gazetteer of Honan (1767 edition)

control developments, also prevented the people from enjoying the fruit of this development. It constituted almost a " permanent " check to the economic growth of the whole region between the Yangtze and the Hwai, despite its natural fertility and wealth.

History has it that " during the period of the Three Kingdoms, the region between the Yangtze and Hwai Rivers was a battlefield, and several hundred *li* were not inhabited. . . . It was only after Wu was conquered that the people returned to their homes."[1] This indicates that the major part of the works of this period in the Hwai valley, as enumerated above, were not only built for military purposes, but had to be maintained by military force, and were therefore subject to the quick changes and wanton destruction that characterize all things military.

The military importance of the Hwai valley, unfortunately, was not limited to the period of the Three Kingdoms. Practically throughout Chinese history since the Ch'in dynasty (255-206 B.C.) the Hwai valley has been the battlefield between North and South, and in the intervals between wars it has always been guarded by military colonization.

" From the beginning of the Tsin to the beginning of the Sui (A.D. 265-589), Hwaiyin was guarded by military colonization; reservoirs and dams were constructed and kept in repair, and grain stored for emergencies. Tsu T'i (A.D. 266-321) protected Hwaiyin with 3,000 soldiers, who also worked on government land. . . . Hsieh Hsüan (A.D. 343-388) first colonized Hwaiyin, then Pei and Hsü; and when armed force and food were sufficient, he took the Fei River and

[1] Shen Yo, *Sung History* (Sung of the Southern and Northern dynasties), *chüan* 35, " Book on Geography," pp. 7-8.

proceeded to Loyang. Tsin also subdued Wu by military colonization of the region north of the Yangtze River . . . and military colonization, together with water-control improvements, was also established in the Hwai region during the Ming (1368-1644) dynasty." [1]

The explanations given by various writers and officials of the military importance of the Hwai valley are illuminating. The Tsin (A.D. 265-420) official, Hsün Hsien (A.D. 321-358), says, "The old *Chen* (fort) of Hwaiyin is geographically a very strategic place. With easy communications by land as well as by water, it is a convenient spot to watch for a chance for conquest. The fertile country can be easily developed and colonized, and boats can cross each other's path in the transport of grain tribute and other products." [2]

A Sung (960-1280) official and scholar, Hsü Tsung-yen, puts it even more categorically. He says, "Shan-yang is a place over which the north and the south must fight. If we [3] hold it, we can advance to capture Shantung, but if the enemy gets it, the south of the Hwai can be lost in the next morning or evening." [4]

Tung Pu-hsiao of the Yüan dynasty (1280-1368) also said: "Hwaian is the throat between the north and south and the strategic point of Kiang and Che (Kiangsu, Anhwei and Chekiang). If it is lost, the Hwai valley cannot be protected. Now, millions of tons of tribute grain are transported yearly through Hwaian. If the throat is blocked, the capital [in the north] faces the danger of immediate death. Hence,

[1] K'ang Chi-t'ien, op. cit., *chüan* 4, p. 45.

[2] Ku Tsu-yü, *A Historian's Notes on Geography* (*Tu-Shih-Fang-Yü-Chi-Yao*), 1774, *chüan* 22, p. 2.

[3] Referring to Southern Sung (1127-1280).

[4] Ku Tsu-yü, op. cit., *chüan* 22, p. 2.

weighty officials are appointed and military colonization established to guard the place." [1]

The secret of this strategic importance was explained by the Sung (960-1280) official, Ch'en Min, who said, "Throughout the 2,000 *li* of the Hwai River, there are five rivers connecting it with the north, the Yin, the Ts'ai, the Ko, the Pien, and the Ssu Rivers; but there is only one avenue southward to the Yangtze, the canal which crosses the Hwai at Ch'u Chou." [2]

Thus its geographic position destined the Hwai valley to the fate of a passageway between the north and south, a centre for military colonization and internal warfare. Under the circumstances, large irrigation constructions were easily reduced to ruins and no enduring constructive development was possible. By being offered on the altar of Mars, the Hwai valley lost the prospect of ever becoming a Key Economic Area, so long as the efforts of men could not change its unenviable position.

SETTLEMENT OF YANGTZE VALLEY DURING THE TSIN DYNASTY

The unity of China under the Western Tsin dynasty (265-317), which succeeded the Three Kingdoms, did not last long. Less than fifty years after its inauguration, the Tsin emperors had to retire to the south of the Yangtze River in the face of victorious rebellions on the part of the peasants in the northern provinces, who were in some districts the descendants of "barbarians" from what is now Chinese Turkistan and Mongolia, who had settled south of the Great Wall

[1] Ku Tsu-yü, op. cit., *chüan* 22, p. 3.

[2] *Ibid.*, Chu Chou is the Sung name for Hwaian.

several centuries before.[1] This period marks the change from Western Tsin to Eastern Tsin (317-420). In the forty-eight years of the Western Tsin, only two instances of irrigation development are recorded in the annals; the repairing of three canals in the Hwai valley in 274,[2] irrigating 1,500 *ch'ing* of land, and the opening up of over ten thousand *ch'ing* of irrigated land along the Chiang and Yü Rivers (in Honan province) in 280.

The transition from Western to Eastern Tsin and subsequent events in the interval before China was unified again under the Sui dynasty in 589 involved a tremendous change in the socio-economic history of the nation. The risings of "barbarian" settlers, who were mostly serfs working on land owned by Chinese "mandarin" lords, as well as the rebellions of discontented Chinese peasants, drove a vast number of Chinese of the upper classes, as well as retinues of their supporters, to the south of the Yangtze River. When the "barbarian" dynasties set up in their northern homes had lasted over a generation, hopes of regaining the northern domain were practically given up in the latter years of the Eastern Tsin, and the Chinese refugees in the lower Yangtze Valley prepared for a permanent stay.

The *Sung Shu* (the history of a minor dynasty, A.D. 420-479) relates the story of the migration in the Hwainan section. It says, "The central domain suffered disturbances, and the *Hu* barbarians frequently invaded the south. Most people in Hwainan, there-

[1] Professor Ellsworth Huntington is historically incorrect in regarding this rebellion of "barbarian" settlers, who had been living in China for several centuries, as a fresh invasion. Huntington, *The Character of Races*, New York, 1924, pp. 182-183.

[2] K'ang Chi-t'ien, op. cit., *chüan* 4, pp. 37-38.

fore, crossed the Yangtze River. At the beginning of the reign of Ch'eng Ti (326-342), Su Chün and Tso Yao revolted in Kiang-Hwai,[1] and the *Hu* invaders again came in large numbers. Thus even more people crossed the Yangtze River [and migrated to the south]." [2]

The conditions which these and other settlers had to face are set forth very clearly in the following account given in the *History of the Sui Dynasty* :

> During the Tsin dynasty, from the time when Yüan Ti moved to the left [3] of the Yangtze River on account of disturbances in the central domain, all those who voluntarily fled to the south were called immigrants. They established districts and prefectures and called them by the place-names of their native land. They scattered and moved about and did not settle at one place. And in the territory south of the Yangtze River [at that time] the custom was to fertilize the land by burning the vegetable overgrowth and, as the seeds were planted, flood the fields with water. The land was low and wet and the people did not save. . . . South of the mountain range,[4] the central government gave official sanction to the authority of local chiefs over their subjects and in return exacted tribute from the chiefs, whose riches consisted of animals, kingfisher feathers, pearls, elephant tusks, and rhinoceros horns. Throughout the various southern dynasties of Sung, Ch'i, Liang, and Ch'en, the system remained unchanged.[5]

It is interesting to compare this account with those quoted above in connection with the study of the kingdom of Wu in the period of the Three Kingdoms.[6] Evidently, the process of colonization in the Yangtze valley was slow, and despite the efforts of the authorities

[1] The region between the Hwai River and the Yangtze Kiang.
[2] Shen Yo, op. cit., *chüan* 35, " Book on Geography," p. 8.
[3] Meaning east, indicating the lower Yangtze valley south of the river.
[4] The Nanling, which separates Kwangtung province from Central China.
[5] Wei Cheng, et. al., *Sui History*, *chüan* 24, " Book of Food and Commodities," p. 3.
[6] Cf. *supra*, pp. 98-104.

of Wu, the Yangtze Valley still remained sparsely settled and, on the whole, very primitive. The external stimulus of a tremendous social upheaval in the north, which was then still the heart of China, intensified by racial and cultural conflict, provided a powerful impetus as well as a pressing necessity for the Chinese people to settle and develop the south.

Such an impetus and necessity for migration had hitherto never been so keenly felt in the history of the Chinese people. The result, as will be seen, was the beginning of a period of rapid development of the fertile Yangtze valley, which ultimately made it the Key Economic Area in China, replacing the Ching-Wei basin and lower Yellow River valley. This brought about a sharp transformation of Chinese culture, and ushered in the crowning maturity of Chinese civilization in the T'ang dynasty.

Water-works in Kiangsu during the Southern Dynasties

At the beginning of this southern migration, in 321, the Tsin minister, Chang K'ai, built the Chua and Hsinfeng Reservoirs, which irrigated 800 *ch'ing* of land. Ch'en Min excavated the Lien Lake in the same year. Both were located near the present Chinkiang district in Kiangsu province, not far from the southern bank of the Yangtze. The Lien Lake assumed considerable importance as a reservoir both for purposes of irrigation and transportation, and was repaired several times in practically every one of the succeeding dynasties. During the T'ang dynasty, any one who secretly cut open the dike of the lake without official permission was punishable by death.[1]

These efforts to improve the conditions of agricultural production in the south of the Yangtze were continued after the Tsin under the four following southern dynasties, the Sung, Ch'i, Liang, and Ch'en. During the reign of Ming Ti (465-472) of the Sung dynasty, the Ch'ihshan Reservoir was built by order of the Emperor.[2] Between 494-497 the Ch'angkang Dam was built by a minister of the Ch'i dynasty.[3] Another Ch'i official petitioned the Emperor to open up 8,554 *ch'ing* of abandoned and new land in Tanyang and neighbouring districts. He estimated that 118,000 workers would be needed to work for one spring in order to construct the necessary dams and tanks. Unfortunately, the plan was not carried out, because the official in charge of the work was ordered to another post.[4]

During the Liang dynasty, in 510, the Hsieh Reservoir was constructed;[5] in 528, an imperial edict was issued ordering one of the officials to facilitate the transportation of grain through the Tatu canal;[6] and in 535, the Chün Canal (now called the Liang Canal) was embanked.[7]

During the Ch'en dynasty, the P'okang Canal was put in order.[8] All of these water-control establishments were located in Southern Kiangsu province, where water-control in later years attained the highest development in China.

[1] *Gazetteer of the Kiangnan Provinces*, *chüan* 64, "Book on Rivers and Canals," pp. 22-27.

[2] *Ibid.*, *chüan* 63, p. 1. In 445, several hundred acres of good land were made productive by the improvement of the Yang Lake. Two years later, the Lichin canal was put in good condition.

[3] *Ibid.*, *chüan* 64, p. 22. [4] *Ibid.*, *chüan* 66, p. 8. [5] *Ibid.*, *chüan* 64, p. 22.

[6] *Ibid.*, *chüan* 63, p. 4. [7] *Ibid.*, *chüan* 64, p. 14. [8] *Ibid.*, *chüan* 63, p. 1.

Forerunners of the Grand Canal

The development of water-routes under the Eastern Tsin dynasty was primarily undertaken for military purposes. During the reign of Mo Ti, from 345 to 361, in order to carry out a military campaign against Mu-yung Lan, who occupied Piench'eng (K'ai-feng), a Tsin official ordered the construction of a canal to lead the water of the Wen River from the Huan River, a tributary of the Wen, to Tunga, in the present Yangku district of Shantung province.[1] About the same time, Ch'en Min, who had made the Lien Lake, opened the Shanyang Yüntao (or Shanyang water-route) from Hsiehyang (near the Hsiehyang Lake) to Mok'ou (near Hwaian in northern Kiangsu); thus the avenue of communication connecting the Hwai River with the Yangtze, which was formerly a very devious route, was reduced to a straight line for the first time.

The importance of these two canals is due to the fact that they were both destined to become sectors of the Grand Canal,[2] which was to play such a significant part in Chinese history in providing a vital link between the seat of political power in the north and the emerging new Key Economic Area in the south under the T'ang, Sung, Yüan, Ming, and Ch'ing dynasties.

[1] K'ang Chi-t'ien, op. cit., *chüan* 4, p. 40.
[2] Cf. *infra*, pp. 113-121 and pp. 139-141.

CHAPTER VII

THE ECONOMIC DOMINATION OF THE YANGTZE VALLEY

THE Yangtze Valley grew in importance as a productive centre during the Eastern Tsin (317-420), and the other southern dynasties (420-589) definitely assuming the position of the Key Economic Area from the time of the T'ang dynasty (618-907). Politically, the centre of gravity still lay in the north. The constant menace of nomadic invasion on the northern frontier emphasized the strategic importance of the northern provinces.[1] Tradition and political inertia undoubtedly also contributed to the decision to keep the political capital in the north, from the Sui (589-618) to the Ch'ing dynasties (1644-1912), despite the shifting of the Key Economic Area to the south.

This anomalous situation rendered the development and maintenance of a transport system linking the productive south with the political north a vital necessity. The link was provided by the Grand Canal, which engaged the attention of the best minds of China for more than ten centuries, and demanded countless millions of lives and a large portion of the wealth of the country for its improvement and maintenance. Its history is closely intertwined with the whole history of irrigation and flood-control and the development of productive areas, and must be studied as an integral part of the unfolding of the whole process.

HISTORY OF THE SUI GRAND CANAL

Although traditionally the canal is ascribed to the genius and extravagance of Yang Ti (605-618) of the

[1] Under the Yüan (Mongol) and Ch'ing (Manchu) dynasties, which were themselves of "barbarian" origin, this factor was, of course, of modified importance.

Sui, it was not built in one period nor by one emperor. Like the Great Wall, it was constructed in disconnected sections at different periods. Yang Ti of the Sui completed it by linking the various waterways running in a north and south direction into a connected system and adding long sectors both in the north and south.

The Grand Canal, thus completed, ran from Ch'an-gan, or Tahsinch'en, utilizing the course of the Wei and Pien Rivers, across Honan province to Hwaian, and then turned southward to Kiangtu (Yangchou) and Kuachou, where it crossed the Yangtze River, and finally to Hangchou. Another sector, which branched out from the Hsin River, a tributary of the Yellow River in northern Honan and southern Shansi, terminated at Chochün, near the present Peiping. The Sui Canal, as a whole, was much longer than the present Grand Canal in mileage, and was a north-south as well as an east-west trunk line of communication. It comprised five distinct sections.

The evolution of the canal can best be studied by examining the history of the different sections. The section with the longest history is the Pien Canal, which bridged the distance between the Yellow River and the Hwai River. The date of the beginning of the ancient original canal, which was called Hung Kou, is unknown, but it must have been earlier than the Spring and Autumn period (722-481 B.C.).[1] It was mentioned by the diplomat statesman, Su Ch'in, of the period of the Warring States (481-255 B.C.). Ssu-ma Ch'ien specifically mentions that after the time of the Great Yü "the Yellow River was led

[1] See Hu Wei, *Notes on Yü-kung* (*Yü-kung Chui-chih*), 1705, *chüan* 5, p. 34.

south-eastward from Yungyang (in Honan province) by [the construction of] the Hung Kou Canal, and connected the feudal states of Sung, Cheng, Ch'en, Tsai, T'sao, and Wei. The Yellow River was thus brought to meet the rivers Ch'i, Ju, Hwai, and Sze."[1]

When Wang Chün (206-285), during the Eastern Tsin dynasty (265-317), campaigned against Wu, the famous Tsin minister, Tu Yü (222-284), in a letter to Wang Chün, referred to his triumphant return with his troops from the Yangtze valley to the capital at Loyang, through the Pien Canal. Hu Wei (1633-1714), the great authority on historical geography in the Ch'ing dynasty, comments on the grandeur of Wang Chün's fleet and army, saying that they were the best and largest of their kind in history, and if they were able to return through the Pien Canal, "the size of the Pien River could not have been smaller than at present.[2] It also shows that this water-route existed during the Ch'in, Han, Wei, and Tsin dynasties, and did not begin with Yang Ti of the Sui dynasty."[3]

However, the old route was different from that of the Sui Canal of Yang Ti. The pre-Sui Canal, according to the able researches of the Japanese scholar, Sadao Aoyama, ran from east of K'aifeng, near the Ch'i district, eastward to Hsüchou, and entered the Hwai River through the Szu River; while the Sui Canal, which was called the T'ungch'i Canal, followed a much shorter and more direct route. It started from Hsi Wan, near Loyang, by leading the Lo and Ku Rivers into the Yellow River; then from Panchu (in the modern Ch'ishui district in Honan province) it branched out from the Yellow River to run eastward

[1] Ssu-ma Ch'ien, op. cit., *chüan* 29, "Book on Rivers and Canals," p. 2

[2] Hu Wei wrote during the reign of Kang Hsi.

[3] Hu Wei, op. cit., *chüan* 5, p. 33.

in its own channel, and on approaching Kaifeng from the east, near the Ch'i district, it turned southward from the course of the old canal and entered the Hwai River directly at Szuchou without going through the Szu River.

Aoyama's conclusion agrees with that of the eminent Chinese scholar, Ku Tsu-yu (1624-1680), author of the famous book, *A Historian's Notes on Geography*. It proves conclusively that the version of the T'ungch'i Canal printed in the *Maps and Gazetteer of Yuanho Chün and Hsien*, the *T'ai-p'ing Yü-lan* and the *Mirrors of History*, which mistook the old route for the new one, was wrong, and that the shorter route, flowing directly into the Hwai without going through the Szu River, as described in the " Biography of Yang Ti " and the " Book on Food and Commodities " in Wei Cheng's *Sui History* and the *Tung Tien*, is correct.

The Pien Canal, or T'ungch'i Canal, which was also called the Yü River, was constructed in 605, and over a million men and women from the prefectures to the south of the Yellow River and north of the Hwai were mobilized for the task.[2] An imperial road was constructed along the bank of the canal and planted with willows.[3] The canal not only combined the navigation facilities of the Yellow River and Hwai River into one system, but wove the tributaries of both rivers into a network of water-routes, covering a section of the North China plain which historically is most important. The economic and military significance of the Pien Canal in Chinese history needs no exaggeration.

[1] Aoyama, Sadao, "Study of the Pien Canal in the Period of T'ang and Sung," the *Toho Gakuho*, or *Journal of Oriental Studies*, Tokyo, No. 2 (December 1931), pp. 1-49, with map showing the routes of both pre-Sui and post-Sui Canals.

[2] Wei Cheng, op. cit., *chüan* 3, " Biography of Yang Ti," p. 5.

[3] *Ibid.*, *chüan* 24, " Book on Food and Commodities," p. 17.

The second section of the Grand Canal, which was called Shanyangtu during the Sui dynasty, ran from Mok'ou, near Hwaian, on the bank of the Hwai River, southward to Kiangtu (Yangchow), thus providing a link between the Hwai River and the Yangtze. The ancient canal bearing the name of Han Kou was dug by King Fu-chai of Wu (495-473 B.C.) in 483 B.C.[1] The original channel ran to the east of the present course, and winding its way through a chain of lakes, especially the Hsiehyang Lake, it was as devious as it was hazardous.

During the reign of Mo Ti (A.D. 345-361) of the Tsin dynasty, the mandarin Ch'en Min cut a more or less straight canal from Hsiehyang to Mok'ou, following a course to the west of the old one, and reducing the distance.[2] It was called the Shanyang Yüntao or Shanyang water-route.

In the time of the Sui dynasty, the canal was silted up, but its course could still be traced.[3] The founder of the Sui Dynasty, Wen Ti, restored the canal in A.D. 587 "to facilitate the transportation of tax grain." [4] In 605, Wen Ti's son, Yang Ti, also ordered improvements of the canal. Over 100,000 inhabitants of the territory south of the Hwai River were mobilized for the work. "From Shanyang (near Hwaian) to the Yangtze River, the water surface of the canal was forty paces wide.[5] Roads were constructed along both banks of the canal and planted with elms and willows. For over 2,000 *li* from the eastern capital (Loyang) to Kiangtu, shadows of trees overlapped each other. A

[1] *Tso Chüan*, *chüan* 58, p. 21.
[2] Cf. *supra*, p. 112.
[3] See Liu Yi-cheng, *A History of Chinese Civilization*, vol. ii, p. 4.
[4] Wei Cheng, op. cit., *chüan* 1, "Biography of Wen Ti," p. 24.
[5] "Dragon boats" were built.

palace was built between every two *Yi* [official post stations], and from the capital in Ch'angan to Kiangtu there were more than forty palaces [built to facilitate the travels of the Emperor]." [1]

The length of the Shanyang Tu was over 300 *li*.[2] Traversing the dividing line between North and South China as it did, the key importance of the canal can be easily appreciated. The territory traversed by the canal was the main battlefield between North and South China.[3] For centuries, before railroads changed the situation, the great strategic value of the city of Hwaian, guarding the northern terminus of the canal, was over and over again demonstrated by history, clearly indicating the crucial position which the canal occupied.

The third section of the canal, known as the Kiangnan Ho or River South of the Yangtze, was dug by the order of Yang Ti in 610. It extended " eight hundred *li* from Ch'inkou (Chinkiang in Kiangsu province) to Yühang (Hangchow in Chekiang province). The water surface measured over ten *chang* (about 100 feet)." [4] This section, which completed the southern terminus of the Grand Canal, enabled Yang Ti and his successors for centuries to tap the wealth of the south-eastern coast of China.

The three sections discussed in the foregoing paragraphs covered the distance from Hangchow to Loyang, which was established as a supplementary

[1] Liu Yü-ching, *Miscellaneous Notes on the Reign of Ta-yeh* (Yang Ti), p. 1. The book was written under the Southern Sung. It is printed as a part of the *T'ang-Sung Ts'ung-shu* or *Repository of the T'ang and Sung Dynasties*. The account in this book is more detailed than the version in Ssu-ma Kuang's *Mirror of History*.

[2] Liu Yi-cheng, op. cit., vol. ii, p. 3, note.

[3] Cf. *supra*, pp. 104-107.

[4] Liu Yü-ching, *Miscellaneous Notes on the Reign of Ta-yeh* (Yang Ti), p. 14.

capital, called the Eastern capital, by Yang Ti. Loyang was linked with the capital, Ch'angan, by the Yellow River and the Wei River, but navigation on the Wei River was often obstructed by heavy " flowing sand, which subjected the river to sudden changes."[1] This difficulty was avoided by the digging of a canal which could use the water of the Wei for navigation. A complete version of the edict which directed the constructon of this canal has been preserved and contains very interesting material :

> [People and goods] from five directions converge at the capital [at Ch'angan] which, surrounded on four sides by mountain passes and fortresses, faces difficulties in communication by land as well as by water. [Happily] the currents and waves of the Yellow River flow eastward and hundreds of streams facilitate communication for ten thousand *li*. Though there are dangers below the Three Gates (Sanmen), they can be avoided if cargoes are transferred on land at Hsiaop'ing ; and after entering Shen (Shensi) by the land route, they could be put back on water and pass from the Yellow River to the Wei River. The Yellow River also dominates the [communications] of the Fen River valley and the territory of Tsin (Shansi). Boats and carts come and go ; the benefit [of easy communication] is certainly great. But the force of the water of the Wei changes often, and when the water becomes shallow and sand thick, it obstructs navigation. The distance this river covers only amounts to a few hundred *li*, but when it changes its condition, boats cannot pass and the labour of the men is interrupted.
>
> My rule over the country is dedicated to the promotion of beneficial things and the removal of the harmful. I regret defects both in the realm of public and private life. Therefore, starting from T'ungkuan in the east and leading the water of the Wei from the west, a canal should be cut by human effort. The work is easy and can be accomplished. Artisans and workers have already looked over the site at my order. Adopting the canal to geographical conditions and considering the meaning [and requirements] of permanence, once the canal is cut, it will not be destroyed for ten thousand generations.

[1] Wei Cheng, op. cit., *chüan* 46, " Biography of Su Hsiao-ch'i," p. 11.

It will enable the government and private persons to navigate big boats, and from dawn to dusk, grain tribute can be transported ceaselessly upstream and downstream. [Thus] the work of several tens of days would save hundreds of millions. I know that in the hot summer, work easily brings fatigue; but without temporary labour, how could permanent rest be made possible? Proclaim this to the people; they should know my wishes." [1]

The canal which ran from Ch'angan to Tungkwan was 300 *li* in length and was named the Kwantun Canal.[2] With the completion of this canal, direct water-route communication was established between Ch'angan and Hangchow.

The fifth section of the Grand Canal branched out from the Hsin River, a tributary of the Yellow River in northern Honan and southern Shansi, and ran northward to Chochün (Cho district in northern Hopei). "Over one million men and women from the various prefectures north of the Yellow River were mobilized by imperial edict to undertake the task." [3] Ssu-ma Kuang (1019-1086), in relating the story, added that "men failed to supply the demand, and women began to be used." [4] The canal was completed in 608 and was called Yungchi Canal. It provided a direct link between the Hai Ho basin (Hopei province) and central Yellow River valley, and the Yangtze valley and Ch'ien-t'ang basin, and afforded a direct means of communication from Hangchow to the vicinity of the present city of Peiping.

[1] Wei Cheng, op. cit., *chüan* 24, "Book on Food and Commodities," p. 14.

[2] It should be recalled that, at the suggestion of Cheng Tang-shih, Wu Ti of the Han dynasty ordered a similar canal to be dug in order to avoid difficulties on the Wei and to shorten a six months' journey to three months. The Han canal was also 300 *li* in length. None of the sources tell whether the Sui Canal followed the course of the Han Canal, but the circumstances seem to point to a similar course.

[3] Wei Cheng, op. cit., *chüan* 3, "Biography of Yang Ti," p. 11.

[4] Ssu-ma Kuang, *Mirror of History*, *chüan* 181, p. 1.

The Waterway System of the Sui, T'ang and Sung Dynasties

In examining the five sections of the Sui Grand Canal as a whole, it is important to note that, except for the two sections from Hangchow to Hwaian which more or less provided the basis of the present Grand Canal, the route from Hwaian northward was entirely different from the present course. The present Grand Canal was primarily the work of the Yüan dynasty (1280-1368) which, for the first time in Chinese history, selected Peiping or Peking as its capital. The Sui Grand Canal was frequently restored and, in places modified, during the T'ang (618-907) and Northern Sung (960-1127) dynasties; but on the whole, the system remained practically the same until the Yüan dynasty. It can properly be called the artificial waterway system of the Sui, T'ang and Sung dynasties.

The Cause of the Unpopularity of Large Public Works Construction in China

Sui Yang Ti's extravagance and cruelty were proverbial in Chinese history, and the Grand Canal was frequently cited as a glaring example illustrating both of these tyrannical qualities. However, these historical facts were generally used by historians, who considered it their thankless task to admonish rulers against becoming tyrants, as lessons in morality. The moral approach of the Chinese historians prevented them from understanding the political meaning of Yang Ti's achievements as well as his crimes.

It is important to realize that, despite the disappearance of a strict caste system in China, the Chinese state was frankly based upon the theory of class rule, and class rule meant the concentraton of surplus resources, very often including a large proportion of the necessities of life squeezed from the people, as an instrument of power and to satisfy the extravagant demands of the ruling group. The concentration of resources demanded canal building and canal building in turn demanded a further concentration of resources, which invariably lead to excessive taxation and a cruel and large-scale programme of forced labour.

A convenient system of communication, following the completion of the canals, naturally stimulated the consuming habits of the ruling groups and further enhanced their extravagance. This inevitable connection between public works development and the increase of exploitation and mass misery explains the great unpopularity of Han Wu Ti and Sui Yang Ti, both very energetic and enterprising monarchs, whose achievements in the field of public works development, including canal building and promotion of irrigation, resulted in public discontent and financial disaster in the one case and loss of the empire in the other.

Had Han Wu Ti not stopped in time by proclaiming his famous " confession " and programme of retrenchment, the rule of the Western Han dynasty probably would have ended a hundred years earlier. Sui Yang Ti, on the other hand, as a clever Chinese critic puts it, " shortened the life of his dynasty by a number of years [by his extravagance in public works construction], but benefited posterity unto ten thousand generations. He ruled without benevolence, but his rule is to be credited with enduring accomplishments." [1] It is only

[1] Yü Shen-hsin, *Dust from the Brush* (A Ming book, 18 *chüans*), as quoted in Fu Tse-huang's *Golden Mirror of the Course of Rivers and Canals*, *chüan* 92, p. 16.

necessary to add that practically all enduring accomplishments of the government in the semi-feudal era in China were tainted with lack of benevolence on the part of the rulers and plenty of misery for the masses.

Large-scale public works, like the construction of the Grand Canal, require a mass-mobilization of labour. In the absence of a developed money economy and a free labour market, this meant forced labour assembled and disciplined by authority of the state. Under the conditions of a semi-feudal class society, this cannot be done without cruelty, and the degree of cruelty necessary is usually in direct proportion to the magnitude of the work involved. The exaggerated degree of cruelty resulting from the gigantic scale of the work on the Grand Canal has particularly attracted the attention of historians. A detailed account of labour conditions in the construction of the Pien Canal, the longest section of the Sui Grand Canal, can be found in *The Record of the Opening of the Canal*, by an anonymous author, which contains the following interesting paragraphs:

> All men between the ages of fifteen and fifty were ordered to assemble by royal edict. All who tried to hide were punishable by decapitation. . . . The labourers thus assembled numbered 3,600,000. Then each family was required to contribute a child or an old man, or woman, to prepare meals for the workers. Five thousand young and brave soldiers were ordered to be armed with sticks [to maintain discipline]. Together with section chiefs and other administrators, the whole number of people employed in the canal amounted to 5,430,000. . . .
>
> At the beginning of the eighth month of the fifth year of Ta-yeh's reign (609 A.D.) baskets and shovels were put to work; the workers were spread over several thousand *li* in a west-east direction. [After a certain sector of the work was done], when the workers were counted, two million and a half labourers and twenty-three thousand soldiers had been lost.

When the work was all done, . . . the Emperor moved from Loyang to Taliang (K'aifeng) and ordered the various prefectures in the Yangtze and Hwai area to construct with the speed of fire five hundred large boats [for imperial use]. Some families among the people to whom had been assigned the duty of contributing one boat, could not fulfill the order, even by sacrificing the whole of their property. They were punished by flogging and neck weights and were forced to sell their children to satisfy official demands. After the dragon boats were completed, they were sent to Taliang (K'aifeng) for decoration. From Taliang to Hwaikou (perhaps, Hwaian) the boats were lined up one after another, for a thousand *li*. Wherever the silken sails passed, their perfume could be smelled for a distance of a hundred *li* ! [1]

The source from which the above quotation is taken, *The Record of the Opening of the Canal*, is considered unreliable by the Ch'ing Imperial Catalogue, because of its "vulgar style," and is regarded as a sort of fictionized history. However, though it may not be accurate in detail, the cruelty of forced labour which it describes is a commonplace of Chinese literature. Even if it gives but a vague indication of the conditions of forced labour, it serves the purpose of contributing to a clearer understanding of the oppressive and unpopular aspect of public works construction in semi-feudal China.

The T'ang Dynasty's Dependence on Kiangnan

However, despite the great suffering which the opening of the Grand Canal must have caused, it provided a link between the two major regional divisions of China, and with its help the capital could successfully tap the resources of the fertile Yangtze valley. Once the Yangtze valley was thus accessible from the capital,

[1] *K'ai-ho Chi* or *Record of the Opening of the Canal*, as quoted in Fu Tse-hung, op. cit., *chüan* 92, pp. 14-15.

a fresh impetus was provided for its rapid development. With its tremendous potentialities released, it soon became the chief producer of grain tribute for the capital, and assumed the position of the Key Economic Area. The *New T'ang History* reveals that

> T'ang established its capital in Ch'angan. But, although Kwanchung was known as a fertile country, the territory was too crowded and its products could not support the capital and accumulate reserves to prepare against flood and famine. Hence grain tribute from the south-east was transported.[1]

Dependence on the south increased to such an extent that by the time of Han Yü (768-824), the great writer, the land tax of Kiangnan[2] had already reached the alarming proportion of nine-tenths of the total land tax of the country.[3]

Small wonder that the task of transporting grain tribute assumed greater and greater importance. From the time when Pei Yao-ch'ing (681-743) made his successful contribution to the problem, it became customary to expect leading officials at court to undertake the responsibility of dealing with the transport of tribute grain.[4] Outstanding administrators of the T'ang regime, such as Liu Yen (715-780), were well known for their achievements in this line.

In comparing the T'ang dynasty (618-907) with the Han dynasty (206 B.C. to A.D. 221), which had struggled primarily with problems of transport in the Yellow River valley, it at once becomes obvious that the T'ang administrators were faced with new technical, social and administrative problems, the solution of which was

[1] Ou-yang Hsiu, *New T'ang History*, *chüan* 53, "Book on Food and Commodities," p. 1.

[2] During the T'ang dynasty, the term Kiangnan included the present Kiangsu, Chekiang, Kiangsi, and Anhwei provinces.

[3] *History Year-book*, vol. iv, p. 95, note.

[4] Ou-yang Hsiu, op. cit., *chüan* 53, p. 3.

one of the most important contributions of the T'ang dynasty to the history of China. The Sui emperors dug the canal, but the dynasty did not endure long enough to solve the other problems of transport. It was the T'ang dynasty which laid the foundation of the grain transport system between the north and south of China. In three years, during P'ei Yao-ch'ing's administration, about 735, seven million tons of grain were transported.[1] The magnitude of the system involved can easily be imagined.

The problem of transportation involves not only the handling of the transport service, but also the maintenance and improvement of the waterways. The two tasks were tackled jointly by the T'ang officials. In the nineteenth year of K'ai Yuan (731), P'ei Yao-ch'ing (681-743) proposed to establish granaries at Hokou and Kunghsien in the present province of Honan, so that " the boats from Kiangnan would not have to enter the mouth of the Lo River." [2] He also suggested to make use of the granaries at half a dozen places along the water-route for the purpose of facilitating transportation, so that when the waterways were navigable the boats could operate, but whenever the water was too shallow the grain could be stored to avoid loss.[3]

The proposal was adopted three years later, in 733, when a granary was established at Hoyin, where the Pien Canal entered the Yellow River. Two granaries were also built along the course of the Yellow River on the southern border of Shansi, one at each side of Sanmen, the " Three Gates," the stone barrier in the bed of the Yellow River which made navigation ex-

[1] Ou-yang Hsiu, op. cit., *chüan* 53, p. 2.
[2] *Ibid.*, " Book on Food and Commodities," p. 2.
[3] *Ibid.*, p. 2.

tremely dangerous. By using the two granaries, boats could be unloaded at the east side of Sanmen and their cargoes transported by land for 18 *li* (6 miles) and stored in the other granary, to await another convoy of boats for further transport. Thus the dangers of Sanmen, the horror of centuries, were avoided.[1] In the same year, Chi Kan, the governor of Yungchow, opened the Yeiko River,[2] reducing the distance of 60 *li* from Yangtzehsin to Kuachou to 25 *li*.[2]

These improvements culminated in the monumental achievement of Liu Yen, who upon his appointment in 764 as Transport Commissioner of the Yangtze and Hwai Rivers, deepened the Pien River, which was then silted up, and adopted with improvements the system suggested by his predecessor, Pei Yao-ch'ing. Special boats were constructed to suit the different conditions and floating capacities of the various sections of the canal. The Kiang or Yangtze boats were to reach Yangchou, the Pien boats to reach Hoyin, and the Ho boats to reach the mouth of the Wei. The Wei boats would finally carry the cargoes to the Ta Ts'ang, the great granary at the capital. Transit granaries were established along the rivers and canals to facilitate changes of boats and to provide for bad navigation conditions which might necessitate waiting.

The system proved to be safe and efficient and with its help and his own genius for administration, Liu Yen attained the record of over thirty years of uninterrupted service, a very remarkable achievement for his time. Thus in Liu Yen's handiwork culminated the system which "later administrators followed." [3]

[1] Ou-yang Hsiu, op. cit., *chüan* 53, p. 2.
[2] Kang Chi-ti'en, op. cit., *chüan* 5, p. 25.
[3] *Ibid.*, p. 26.

Rise of Water-control Works during the T'ang Dynasty

The data in Chapter III indicate the sudden rise of irrigation activities in the T'ang dynasty in practically all the provinces except Honan. To follow the method which has been followed hitherto, of enumerating all the major works constructed, would be superfluous. Ancient irrigation and transport works were fewer in number and were of greater historical importance in shaping the development of the system; hence the different works have been separately enumerated above and the story of the construction of some of them told. Many of the T'ang works were reconstructions of old systems. New works, however, particularly in the south, were numerous, but for the purpose of the present investigation need not be treated individually and in detail.

A feature of the history of the T'ang works which deserves special attention is the fact that according to the great Ch'ing scholar, Ku Yen-wu (1613-1682), seven-tenths of the water-control works recorded in the "Book on Geography" in the *New T'ang History* were constructed before the reign of T'ien Pao (742-756).[1] When it is recalled that T'ien Pao was the reign-style of Hsuan Tsung, after the rebellion of An Lu-shan which plunged the country into years of civil war and destruction, and reduced the financial power and prestige of the T'ang dynasty to a fractional part of its former dimensions, the decline in water-control activities after that period is easily understandable. Ku Yen-wu was right in suggesting that perhaps, after

[1] Ku Yen-wu, *Notes on Daily Studies* (*Jih-chih Lu*), *chüan* 12, p. 25.

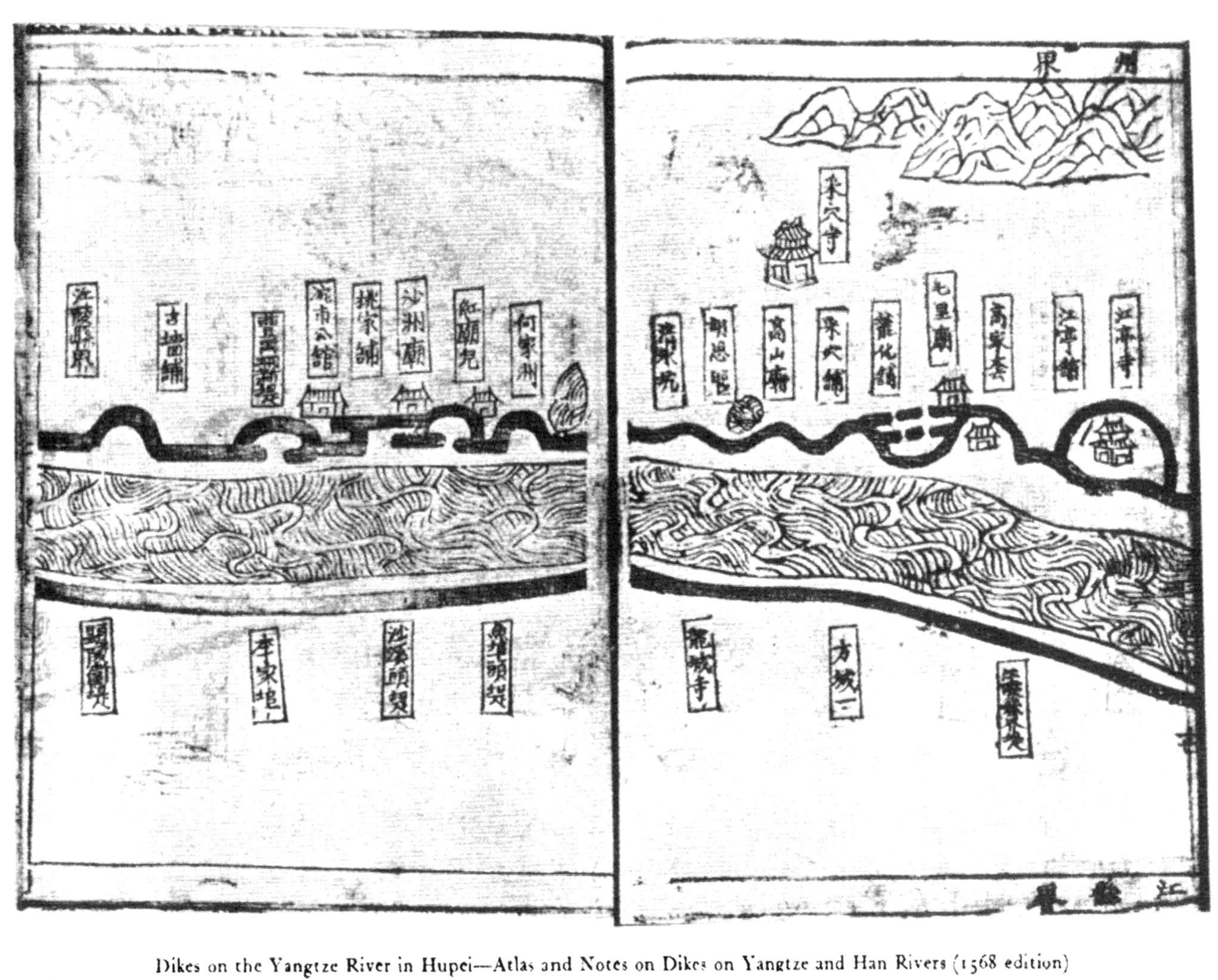

Dikes on the Yangtze River in Hupei—Atlas and Notes on Dikes on Yangtze and Han Rivers (1568 edition)

the war, the government, busily engaged in pushing the collection of taxes, had no time to devote to such constructive labour as the encouragement of agriculture and the improvement of waterways.[1]

NEGLECT OF WATER-CONTROL IN THE NORTH

The swampy jungle-covered Yangtze valley of ancient China must have offered infinitely greater obstacles to the early settlers than the northern steppes and Huangho or Yellow River delta. But once development began on a large scale, as it did after the great migratory movement of the Eastern Tsin (265-317), the potentialities of the region were so great and the return on labour and capital so tempting, that it easily began to outstrip the older and more civilized North.

During the Sui and T'ang dynasties the rate of progress was greatly enhanced by the construction of the Grand Canal, and the South at this period definitely caught up with the North. The North, however, was not entirely neglected at first. The fact that the T'ang arose as a northern dynasty, and that the political inertia which caused the North to be looked on as the centre of gravity of Chinese life was still strong during the T'ang Dynasty, prevented the North from falling in disfavour for some time, despite the dominant economic position which the South had already attained.

The result of the continued care for the North can be seen from an edict issued in 720, which describes Kuanchung as a country where "fertile land spreads before one's view and ditches and canals run into each

[1] Ku Yen-wu, op. cit., *chüan* 12, p. 25.

other; where grain is stored in the granaries, and the capital area yields irrigation benefits worth an ounce of gold for each *mu*."[1] But with the decline of the T'ang and the disturbances of the Five Dynasties (907-960), accompanied by great devastation, particularly in the northern provinces, signs of neglect began to appear. As the Sung dynasty emerged from the struggles in 960 and reunited China, the shift of emphasis became even more evident. This tendency can be clearly seen from the table in Chapter III, from which the following figures of public works undertakings in the major provinces are taken:

NORTHERN PROVINCES

	Shensi	Honan	Shansi	Chihli
T'ang . . .	32	11	32	24
Northern Sung . .	12	7	25	20
Kin, contemporaneous with Southern Sung .	4	2	14	4

SOUTHERN PROVINCES

	Kiangsu	Chekiang	Kiangsi	Fukien
T'ang . . .	18	44	20	29
Northern Sung . .	43	86	18	45
Southern Sung . .	74	185	36	63

It is significant that even under the Northern Sung, when the Sung dynasty still ruled the whole of China (960-1127), more attention was paid to the development of the South than that of the North.[2]

[1] Miao Feng-ling, *Outlines of Chinese History*, Nanking, 1932, vol. i, p. 68.

[2] See also Liu Yi-cheng, *History of Chinese Civilization*, vol. ii, pp. 218-219. Liao (907-1125) and Kin (1115-1260) ruled the northern provinces while Southern Sung (1127-1280) was established in the south. No data of irrigation works during the Liao dynasty are available.

SUB-REGIONAL DIVISIONS IN THE YANGTZE VALLEY DURING THE PERIOD OF THE FIVE DYNASTIES

But the Yangtze Valley during the T'ang, although contributing a large proportion of grain tribute, had not yet reached a mature stage of development as a consolidated economic area. The topographical character of the country is indicated by the remark of Professor Cressey that "all of China south of the Yangtze is a land of hills and mountains."[1]

These hills and mountains cut the region into six distinct sub-regions which made unity difficult under the level of economic development at that time. Barring Szechwan and Yunnan and the two Kwangs (Kwangtung and Kwangsi), whose character as economically self-sufficient and independent regions persisted to a much later date, the four areas that were practically independent regional units were natural geographical regions.

Each of the four regions provided a base for an independent state following the dismemberment of the T'ang dynasty. The T'aihu and Chient'ang valley (present Chekiang and southern Kiangsu) was occupied by Wu Yüeh (907-977). The Hwai valley and lower Yangtze valley (present northern Kiangsu, Anhuei and Kiangsi) was the seat of the dynasty of Wu (907-937), which was later succeeded by Nan-T'ang (937-975). Fukien was for some time the territory of the state of Min (907-945), while the present Hunan province and part of Hupei once comprised the state of Chi (907-922).[2]

[1] George Babcock Cressey, op. cit., p. 41.

[2] Compare the five maps on pp. 29-31 in Ou-yang Yin's *Historical Atlas of China*, Wuchang, 1933.

Thus, while practically the whole Yellow River valley was dominated by the " Five Dynasties," in succession (907-960), the Yangtze valley, though economically greatly developed during the T'ang dynasty by water-control undertakings, had not yet advanced far enough to overcome natural and historical barriers between the various independent and self-sufficient constituent units.

In other words, the Yangtze valley, which had contributed a large proportion of the grain tribute under the T'ang dynasty, had, as a whole, in contrast with the Huang Ho or Yellow River valley, assumed the position of the Key Economic Area ; but the history of the fifty-three years of the Five Dynasties indicates that the Yangtze valley was still composed of very loose units which had not yet grown into one closely knit homogeneous area. During the period of the Three Kingdoms (221-265), the Yangtze valley had been still too superficially touched by civilization to be territorially differentiated into large, distinct and powerful component areas. After the Sung dynasty, which ended with the Mongol Conquest in 1280, the whole valley, except Szechwan, was sufficiently developed, in regard to communications, as well as cultural homogeneity to allow the major part of the territory to be woven into one regional unit. It is significant that since the Sung dynasty there has been no sustained division in China.

The two outstanding periods of sustained division in Chinese history not primarily caused by invasion were the period of the Three Kingdoms and the period of the Five Dynasties. To state the problem from the standpoint of the economic base, or the Key Economic Area, the former period was due to the rise of rival areas which weakened the relative economic supremacy of the dominant area, while the latter was the manifestation of the fact that the most productive area had not yet become one unit, so that its internal differentiation weakened its potential power.

Thus, during the latter period, the less productive but more homogeneous and organized area, the Huang Ho or Yellow River valley, maintained its dominant position, though it failed for decades to subjugate the South. The Posterior Chou (951-960), the last of the Five Dynasties which occupied the Huang Ho valley, reinforced its economic position by many constructive enterprises, especially water-control works.[1] These achievements strengthened the economic and therefore the military power of the man who succeeded the Posterior Chou dynasty, Chao K'ang-yin, and helped him to bring China once more under one rule and to found the dynasty of Sung.

But the Sung did not persist very long as a united power in China. Wave after wave of Tatar invasion drove the Sung power southward. This contributed greatly to the completion of the development of the Yangtze valley under the technical resources then available. The Yangtze valley toward the end of the Southern Sung was essentially the same as the Yangtze valley on the eve of modern industrialization.

Development of Lake-bottom Land under the Southern Sung

" After the Sung dynasty crossed the Yangtze valley to settle in the south, the productivity of irrigated land was found to be richer than in the central domain ;

[1] K'ang Chi-t'ien, op. cit., *chüan* 5, pp. 48-56.

therefore, activities for making use of water-benefits were greatly increased."[1] In these few words, recorded in the *Sung History*, can be seen another epoch-making step in the opening up of the Yangtze valley. The pressure from the Liao (907-1125) and Kin (1115-1260) invaders forced the Sung dynasty and a large proportion of its subjects to move bag and baggage to the south.

The history of the Sung dynasty is replete with records of water-control developments. The court and officialdom seem to be more concerned with irrigation than did those of any previous dynasty. The extensiveness of the development can be seen from one of the recorded cases. In 1174, for instance, an official reported the completion of 24,451 works for water-control, irrigating 44,242 *ch'ing* (about 752,000 acres) of land in the forty-three *hsien* (district) under his jurisdiction.[2]

With its new capital established at Hangchow, the unprecedented increase in population stimulated agricultural productivity which, in turn, lead to a demand for more and more land. Under the geographical conditions of South China the solution of this problem was sought in the drainage of the marshes and even some of the lakes. The result was the development of the so-called Wei-land or Yü-land, which are together designated here as "lake or river bottom land," or briefly as lake-bottom land, a unique feature in South China agriculture.

Lake-bottom land began to attract attention by the reign of Cheng Ho (1111-1117). The reclamation by drainage of large sections of lake bottom and river

[1] T'o T'o, *Sung History*, *chüan* 173, "Book on Food and Commodities," p. 29.

[2] Pi Yüan, *Continuation of the Mirror of History*, *chüan* 144, p. 3.

bed, with dikes to keep the water out, added considerably to the cultivable area under the Sung dynasty. Since such land was usually lower than the surrounding water level, any defect in the dikes would cause flooding and great damage. Hence constant repair of the dikes became a vital necessity.

Many repair works and constructions of new *Yü* and other water-control works done by imperial edict during the Southern Sung (1127-1280) testify to their importance. Even during the later dynasties the *Yü* have remained so much an integral part of the agricultural system that a heavy responsibility fell upon the shoulders of the peasants, and especially the magistrates, in keeping them in order. When the *Yü* went out of order, not only did the peasants face starvation but government finances also were menaced.

Wei-t'ien : A grave Socio-economic Problem

But the practice of reclaiming land from lake bottoms and river beds created a serious socio-economic problem. The absolute dependence of agricultural production on the water-control functions of a government which, under the system of semi-feudal private ownership, was dominated by semi-feudal landed interests, gave rise to deep-rooted contradictions.

On the one hand, the extraordinary fertility of lake-bottom land tempted landowners and peasants to extend their cultivation. On the other hand, the danger of encroaching too much on the area necessary for drainage, a danger which could only be avoided by strict governmental control, remained unchecked because the nature of the government was such that it was incapable of working out a rational system that would enable it to discharge its functions properly. Furthermore, under the guise of military colonization, officials used to occupy the choicest land, sometimes blocking the most vital drainage lake or stream. The consequence of the whole situation is described by Wei Chin, an official of the Southern Sung dynasty. He says:

Since the last years of the reign of Shao Hsing (1131-1162), the usurpation of marsh land inside the shores of the lake[1] by soldiers has led to the construction of dikes that brought into being the so-called Pa-land, similar to Wei-land, from which the people have already suffered bad consequences. After the reigns of Lung Hsing (1163-1164) and Chein Tao (1165-1173) not a single year passed without the powerful clans and influential families following one after another in the occupation of large areas of lake-bottom land.

Thus the people were deprived more and more of the benefits of reservoirs and lakes. What used to be rivers, lakes, and marshes has all become land in the last thirty years. It was the powerful families, whose words and influence could dominate the government, that created the Wei-land. And those who were in positions of responsibility were over-cautious and indulged in passivity and procrastination. The superiors (officials) and inferiors (powerful families) shut their eyes to each other's acts and pretend not to know that anything has happened.

But the harm done by the Wei-land has become serious. There are those who argue that if Wei-land becomes widespread, tax increases will be large and may therefore help the government's finances. Those who talk this way do not realize that hundreds of thousands of [*mu* of] good land belonging to the people lie along the rivers and lakes, and none of them suffered flood or drought at other times. The rise of Wei-land and the construction of dikes and embankments have blocked the outlet of the waters. During slight drought, the owners of Wei-land take possession of the upstream sections of the rivers and monopolize the advantages of irrigation, and the people's land is deprived of the use of water. When the water of the rivers and lakes overflows, the surplus water is sent downstream and the people's land is used as a

[1] Probably the Tai-Hu.

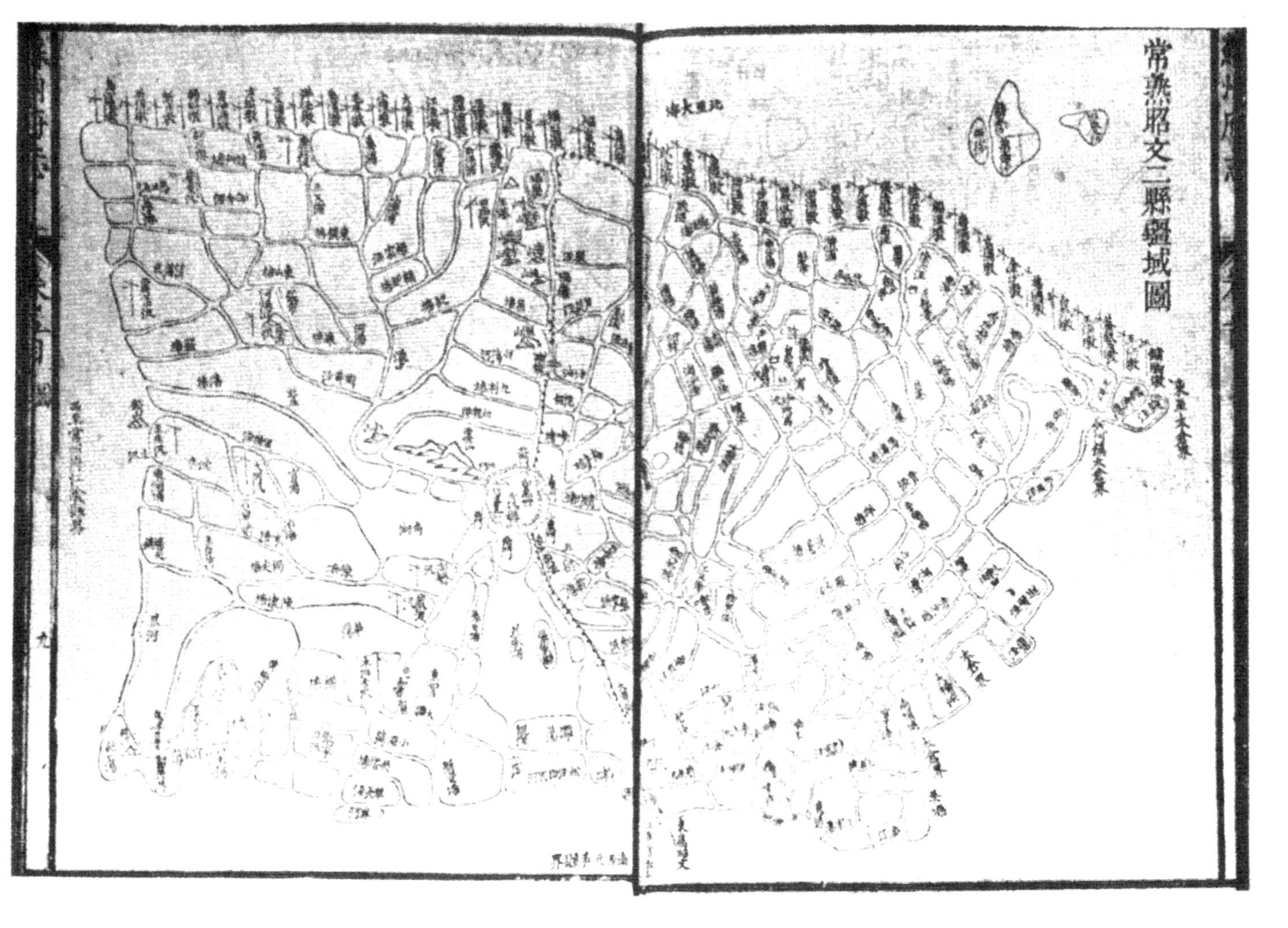

Typical Irrigation and Drainage System in Lower Yangtze Delta (Ch'angshu and Chaowen Districts, Kiangsu)
—Gazetteer of Suchou (1881 edition)

water-shed. Even if fortune favours the Wei-land with a good harvest, and rent increases and taxes are doubly collected, the people's land is reduced to waste whenever there is flood or drought. The damage thus done to the normal tax income is indeed incalculable ! [1]

Scores of similar documents and many more references to the same problem can be found in the histories and other literature of the Sung period, as well as the records of subsequent dynasties. When the damage done to the treasury became so alarming that the court was forced to take steps to ameliorate the situation, edicts prohibiting the diking up of lake bottoms and river beds were issued, but to no avail. When, in 1183, an order was given to inscribe such an edict on stone tablets, one of which was placed on every tract of Wei-land, 1,495 tablets were required.[2] But there was little improvement, and in 1196 the situation was described by Yüan Shu-yü of the Department of Public Works, in a memorial to the throne, as follows :

Tracts of Wei-land lie one beside the other covering a territory of hundreds of thousands of *mu* in Chehsi (Chekiang). Tanks, ponds, reservoirs, streams, and creeks are all turned into farms. No trace of any storage for surplus water can be found and no water to relieve drought is available. If not strictly prohibited, the situation will become worse, and years of good harvest can hardly be expected ! [3]

The reason for the impotence of the government in dealing with the matter lay in the fact that the offenders were actually the most powerful personages in the government. This was not only true of the local government, but also true with regard to the

[1] *The Imperial Encyclopedia of Agriculture* (*Ch'in-ting Shao-shih T'ung-k'ao*), *chüan* 12, pp. 12-13.
[2] T'o T'o, op. cit., *chüan* 173, "Book on Food and Commodities," p. 35.
[3] *Ibid.*

central government. The Sung scholar, Ma Twan-lin, indignantly refers to the fact that east of the Yangtze, Tsai Ching and Ch'in Kuei, notorious leading ministers at the Sung court, successively owned the Wei-lands. He continues to the effect that

> The lands of to-day are mostly the lakes of yesterday. Those who are responsible for the Wei-land seem only to know that the lakes can be drained and reclaimed for cultivation, but do not seem to realize that land outside the lake will thus be flooded. This is because the responsible parties are court favourites and powerful officials; hence they can, without fearing any interference, condemn the neighbouring land to the fate of a water-shed and thus benefit themselves by hurting the people.[1]

However, such a situation does not necessarily have to accompany the utilization of fertile lake bottoms and river beds for agriculture. Trouble arises when a government is absolutely incapable of discharging the functions of a collective agent in a region where irrigation agriculture makes close co-operation and centralized collective planning and administration an absolute necessity. Had such a centralized collective planning and administrative agency been forthcoming in China, the needs of lake and river drainage could have been met with the use of a minimum of space, and perhaps even a larger area than has hitherto been possible could have been reclaimed for agriculture. The whole problem thus comes face to face with the land system and the nature of the state.

This has been the central socio-economic problem in the Yangtze valley ever since it reached full maturity, under the Sung dynasty, as the Key Economic Area in China. In our own time, the solution of this funda-

[1] Liu K'e-yu, *The Nine Encyclopedias of Source Material on the Institutional History of China (Chiu-t'ung T'ung), chüan* 9, p. 4.

mental problem, in conjunction with the problem of land ownership and political power, still constitutes one of the major tasks of the Chinese people. The reverberations of the various manifestations of the problem which affect the productivity of the Key Economic Area cannot but be felt throughout the whole domain of China.

The Yüan Grand Canal

Beset by this and other grave problems that could find no solution, the Southern Sung dynasty gradually weakened under the increasing burden of their cumulative effects, and finally succumbed to the Tatar invaders who founded the Yüan or Mongol dynasty (1280-1368) and again united China under one rule.

For reasons of politics, the Mongols selected Peiping, which they called Tatu, the Great Capital, a strategic fortress near their home, as the capital. But they soon realized that if they were to remain long in the seat of power in China they must rely upon the Yangtze valley, the Key Economic Area, as the base of supplies.

But the Sui Canal, which linked Chochün, near Peiping, with the Yangtze valley, followed a devious course that extended far to the west in the valley of the Hsin River, before turning eastward again to meet the north-south canal from Hwaian to Yangchou. The course of the Sui Canal had been designed to serve the capital at Ch'angan during the Sui and T'ang dynasties. It had also been adequate to the needs of the Northern Sung, when its capital was at K'aifeng. But when the Yüan dynasty located its capital at Tatu in the valley of the Hai Ho, the course of the canal was no longer convenient.

Furthermore, the Southern Sung army had deliberately destroyed the northern section of the canal in 1128, by cutting open the embankment of the Yellow River.[1] In view of this and the other damage it had suffered during the long period of armed struggle between the Southern Sung and the Tatar dynasties in the north, the Sui Canal must have been in a very dilapidated condition when the Yüan dynasty came to power.

Although ocean transport was of great importance during certain periods in the Yüan dynasty, the risks to human life, and the property losses, were tremendous. The Yüan dynasty was still forced to rely chiefly on inland canal transport. Hence the reconstruction of the Grand Canal became one of the major tasks to be dealt with by its administrative machinery.

The Yüan canal, which was used by the Ming and Ch'ing dynasties and is still in existence at the present day, consisted of six sections. The first section, called the T'unghui Canal, ran from the capital to T'ungchou, and in its construction 19,228 soldiers, 542 building artisans, 319 water experts (shiu-shou), and 172 criminals were engaged from the spring of 1292 to the autumn of the following year, under the direction of the famous water-control authority, Kuo Shouching. The finished canal was 170 *li* and 104 *pu* (paces) long.[2]

The section called the Po Ho, connecting T'ungchou with Chihku (north of Tientsin), was made serviceable for grain transport by frequent improvement and deepening during the Yüan dynasty. The third section, called the Yü Ho, or Imperial River,

[1] K'ang Chi-ti'en, op. cit., *chüan* 6, p. 79.

[2] K'e Shao-ming, *New Yüan History*, *chüan* 53, "Book on Rivers and Canals," pp. 1-2.

which extended from Chihku to Linch'ing, was really a section of the Wei River improved and made navigable. Its southern terminus ran into the Huit'ung Canal, the fourth section of the Yüan Canal, which ran from the Hsüch'en district in Tungch'ang (in Shantung province) to Linching. The Huit'ung Canal was fed by the Wen River. Dug in 1289, it was 250 *li* long and had 31 locks; 2,510,748 work-days[1] were spent in its construction.[2]

The fifth section was called the Yangchou Grand Canal and ran southward from the Huit'ung Canal at Shantsako (in Linhsien in Shantung province). The sixth section was the so-called Chinkiang (Chengchiang) Canal, running from Chinkiang to the Lüchen Dam in Ch'angchou. The southern sections followed the route of the Sui Canal more or less closely. The labour spent during the Yüan and later dynasties on this section was primarily for maintenance and improvement.

CANAL MAINTENANCE AND FLOOD-CONTROL ON THE YELLOW RIVER

The Yüan Canal served as the main avenue of communication between north and south during the Yüan and Ming dynasties and also the Ch'ing dynasty. The main problems facing the authorities of the three dynasties were problems created by the fact that one sector of the Grand Canal merged with the Yellow River.

The close relation between the maintenance of Grand

[1] One full day's labour by one man, usually working from sunrise to sundown, was counted as a work-day. This method of counting the magnitude of a specific piece of labour is still prevalent in China to-day.

[2] K'e Shao-ming, op. cit., *chüan* 53, pp. 8-9.

Canal transport and flood-control of the Yellow River greatly emphasized the importance of both, and elevated the administrative officials in charge of both tasks to very high positions in the bureaucratic hierarchy. The close relation between the two tasks was clearly set forth by P'an Chi-hsün, the great Ming official expert, who says:

> Great benefit to the Grand Canal transport system can be gained by proper care of the Yellow River. Since the transport of grain tribute cannot be stopped for a single year, the Yellow River should not be neglected for a single year. To attain two ends by one act, the control of the Yellow River is indeed important. Thus, before the Sung and Yüan dynasties, the Yellow River changed its course often suddenly to the north and suddenly to the south, without a single year of peace. But under the present regime, it has not yet changed its course during more than two hundred years. This is due to the care it has been getting as a sector of the Grand Canal.[1]

In this connection, it is important to point out that throughout the semi-feudal epoch in China, the government always considered the interest of grain transport above that of irrigation or flood-control. The former was primarily an act of appropriation, identified with the immediate enjoyment of the fruit of rulership and the obvious necessity for the maintenance of power by armed force, fed by the grain tribute. Irrigation and flood-control were more immediately a question of peasant welfare, and were more remotely, though vitally, connected with appropriation and the question of power.

The short-sightedness of the semi-feudal bureaucracy, perhaps an inevitable result of its position, can be seen more clearly from a memorial to the throne

[1] P'an Chi-hsün, *Handbook on River Control*, *chüan* 3, p. 36.

submitted by P'an Chi-hsün in 1588. In this memorial, he revealed that the bank of the Yellow River, upstream, near Shanmen and Tseching, could easily be burst, but "because it is not used for navigation, it is usually neglected."[1] P'an criticized this neglect, not because it brought misery on the heads of millions in Honan province, but because "if the bank of the river upstream ever burst the transport canal downstream would certainly be blocked."

The misery of millions, so long as it had not fermented into rebellion, never touched the heart of the rulers in the same way as the urgency of fiscal appropriation and the necessity for maintaining the right of appropriation. This sad truth lies behind the orientation of flood-control policy that has here been described. This is the law of power politics, the law that determined the course of water-control development and governed the shifting of Key Economic Areas in Chinese history.

Attempt to Develop the Hai Ho Valley into a Key Economic Area

The greatest achievement of the Yüan dynasty in water-control was the Yüan Grand Canal. The data in Chapter III indicate that its record in such matters, as a whole, lies far behind that of the Ming and Ch'ing dynasties. The Ming dynasty owed its good start in this work to its first Emperor, T'ai Tsu, who at the beginning of his reign ordered that all petitions in regard to "water-benefits," from the people as well as from officials, be brought to his attention as soon as they were received.

[1] *Gazetteer of Honan*, *chüan* 14, p. 23.

In the twenty-seventh year of his reign (1394), he ordered the Department of Works to make necessary repairs on reservoirs, ponds, lakes and dikes, and advanced students from the national educational institution in the capital, together with special experts, were despatched to the provinces to supervise the work. In the following winter, reports came from all parts of the domain to the effect that a total of 40,987 water-control works had been completed.[1]

The Ch'ing dynasty also holds a good record in water-control activities. During the reigns of K'ang Hsi (1662-1722), Yung Cheng (1723-1735) and Ch'ien Lung (1736-1795), a great deal of attention was paid to irrigation and flood-control works. Most of the activities during these three dynasties were centred in the Yangtze valley, the chief provider of tribute grain, to the neglect of the northern provinces.

But not all northern provinces were condemned to the same fate. Chipu (the metropolitan area) in the Hai Ho valley, the seat of the capital, received special treatment and contributed a very interesting chapter to the history of water-control in China. The reason for this special treatment was two-fold. One of the reasons was stated by Hsü Cheng-ming (*d.* 1590), a Ming official, who in a petition suggesting water-control improvements in the capital area expressed the opinion that food for the support of the army in the capital was a weighty consideration, and that it "should be sufficiently provided for in the capital area. There is great danger in the present situation," he says, "in the almost complete dependence on the grain tribute from the south-east, without which

[1] Chang Ting-yü, *Ming History*, *chüan* 88, "Book on Rivers and Canals," p. 1.

the army would be unable to get along for a single day."[1]

The second reason was most clearly brought out by the famous Ch'ing mandarin, Lin Tse-hsü (1785-1850), the Commissioner Lin of the "Opium War" at Canton in 1840. In a memorial to the throne, he argued that "with the seat of the capital in the north, and transport of grain from the south for its support, each *tan* of grain in the capital granaries usually costs the value of several *tan* for its transport. Although long experience shows that the transport can be indefinitely continued, yet, to manage the country with a far-sighted policy and to plan for ten thousand years to come, further improvements can still be made."[2] In the same memorial, he calculated that if 20,000 *ch'ing* of land were added by reclamation in Chihli (now Hopei), the capital area would be able to produce the four million *tan* of grain tribute annually transported from the south.[3]

Thus, with a view to saving transportation costs and making the capital self-sufficient in regard to food, particularly for the army, officials of the Yüan, Ming and Ch'ing dynasties repeatedly suggested schemes for converting Chihli into a "second Kiangnan," or, in other words, for creating a Key Economic Area near the seat of the capital.

In 1352, at the suggestion of the premier, T'o T'o, the Yüan emperor ordered that all government land, and land formerly used for military colonization in the Hai Ho valley, from the Western Hills in the west to Ch'ienmingcheng in the east, and from Paoting and Hochien in the south to Tanshun in the north, be

[1] Hsu Kuang-chi, *Encyclopedia of Agriculture*, *chüan* 12, p. 8.
[2] *Gazetteer of Chipu*, *chüan* 91, p. 71.
[3] *Ibid.*

distributed to tenants to be cultivated on a share-crop basis. At the beginning of the settlement wages, oxen, farm implements and seeds were to be provided by the government. About one thousand farmers proficient in the cultivation of irrigated land and one thousand experts in irrigation work construction were hired from the Yangtze provinces to act as instructors to the peasant-tenants. The *Yüan History* records that as a result of these efforts the region enjoyed good harvests yearly.[1]

Similar attempts and suggestions were made by many officials of the Yüan, Ming and Ch'ing dynasties. It would be superfluous for the present purpose to go into details. They all follow similar lines of argument, and the outlines of the various proposals, and the results in practice, were comparable. The work of T'o T'o, one of the earliest attempts, provided a sort of model, and was frequently referred to in later discussions.

The immediate results of the series of attempts to develop the Hai Ho valley, however, were not very encouraging. Good harvests were occasionally reported as a consequence of the improvements, but the hope which the promoters cherished, of being able to use Chihli products as a substitute for the grain tribute from Kiangnan, thus making grain transport from the south unnecessary, was never realized.

The Yangtze Valley as the Key Economic Area under the Yüan, Ming and Ch'ing Dynasties

The lower Yangtze valley had already attained the status of being the Key Economic Area during the

[1] *Gazetteer of Chipu, chüan* 90, p. 5.

T'ang dynasty (618-907), but it had not yet assumed decisive political importance, because of its internal sub-regional looseness. But it emerged from the Southern Sung dynasty (1127-1280) with an internal cohesiveness that grew in solidarity with its further development during the following three dynasties.

The Yüan or Mongol dynasty (1280-1368) ruled China from the north as an invader's regime, but throughout the dynasty, the ruling house manifested a certain fear of the wealth of the south and was alarmed over its potentialities. The policies adopted to allay this fear were the improvement of the grain transport system and the attempt to convert Chihli into a second Kiangnan.

By the time Chu Yüan-chang, the founder of the Ming dynasty, rose to conquer the country from the Hwai valley, he already found his strongest enemy located in the Yangtze valley. Hence, when he had succeeded in unifying the country again under one rule, he selected Chinling (Nanking) as his capital. The capital was again moved to the north by Cheng Tsu for purely political reasons, and thenceforth the Ming dynasty was subject to the same worries that had perplexed the Yüan, about the great distance separating its political base from its economic base. The Ch'ing or Manchu dynasty selected Peking as its capital for the same reasons that had motivated the Mongols who founded the Yüan dynasty, and also worried over the same vital problems throughout its regime.

The attraction of the Yangtze valley as the centre of gravity, economically as well as politically, was also expressed in the growing importance of the city of Nanking, the chief city of the region. The key economic position of Nanking was very shrewdly described by a literary adherent of the T'aip'ing Rebellion (1851-1863) in the following passage:

> Hupei, Honan and Chinling [Nanking] all occupy central positions under heaven [in China]; but Hupei and Honan both suffer from the menace of flood, while Chinling is not only geographically situated on an elevation, but the people are rich and prosperous. Furthermore, the world's grain is all produced in the south. It would be convenient indeed to transport grain downstream from Kiangsi and Anching. Chekiang and Kiangsu are even nearer, and can be speedily reached. Thus the ten thousand countries can easily assemble in this centralized location, which should be selected as the seat of the capital.[1]

These words were written at a time when Nanking was truly the leading city in the region, before the changing nature and role of the Key Economic Area reduced it to the humiliating position of a political annex of Shanghai. The position of the Yangtze valley as the Key Economic Area, with Nanking as its leading city, was destroyed by the breaking down of Chinese isolation in the middle of the last century and the beginning of a new era in Chinese history.

General Conclusion—The Changing Nature of the Key Economic Area

As a conceptional device for the study of regional relations, the idea of the Key Economic Area so far has contributed to the illumination of the character of the regional relations in the semi-feudal epoch in China. It was an epoch in which agriculture, especially irrigation farming, was the leading industry, in which agricultural production was dependent upon the proper

[1] Shen Shih-chu, "Essay on Chinling as the Capital of the Empire," printed in *Source Material on the History of the T'aip'ing Rebellion*, first series, vol. ii, p. 9.

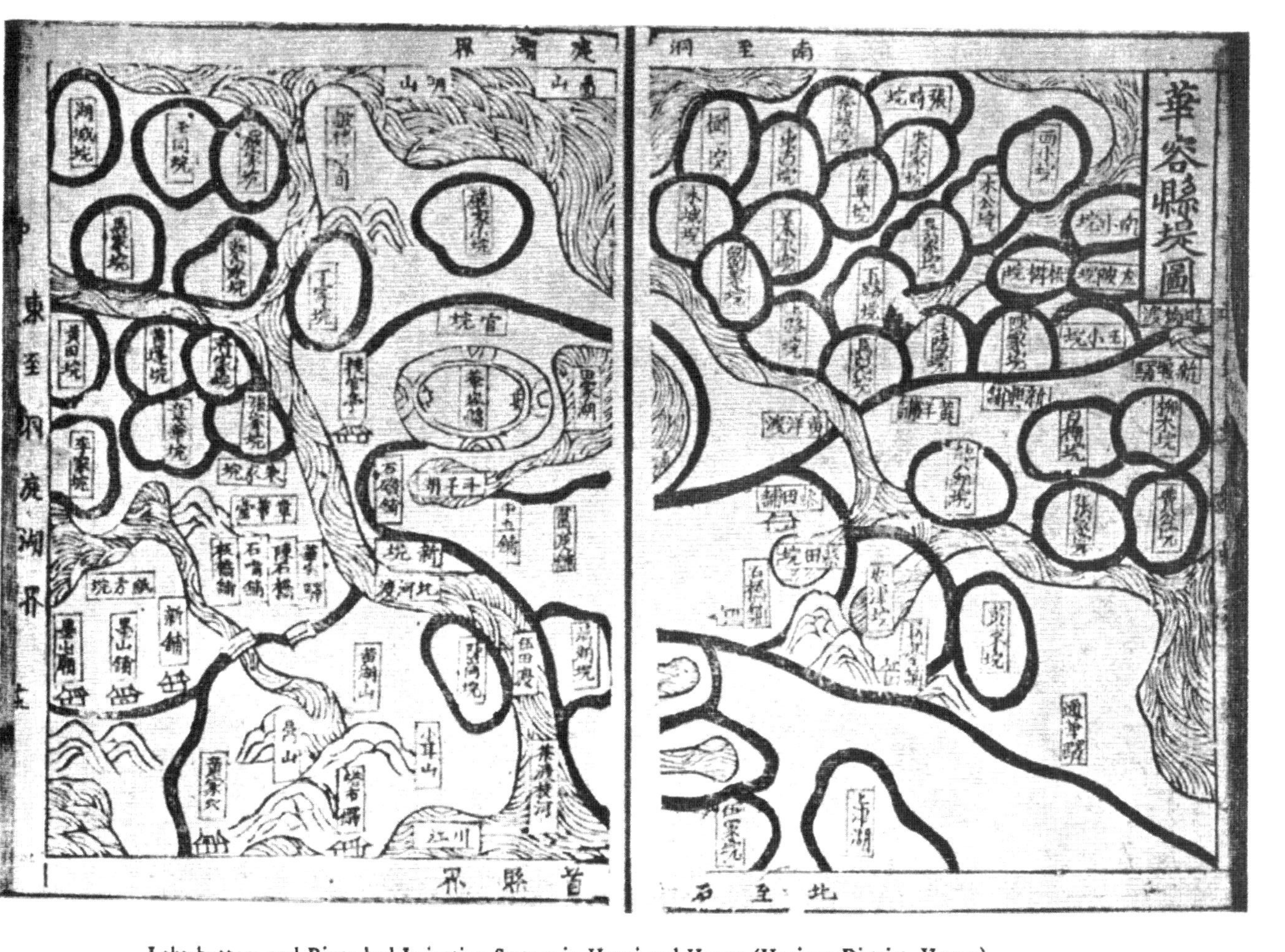

Lake-bottom and River-bed Irrigation System in Hupei and Hunan (Huajung District, Hunan)
—Atlas and Notes on Dikes on Yangtze and Han Rivers (1568 edition)

functioning of a great variety of water-control works constructed and maintained by the state, and in which the state, dominated by a landowning bureaucracy, used water-control activities as an economic weapon of political struggle and a chief means to develop and maintain a Key Economic Area as an economic base for the unified control of a group of more or less independent self-sufficient regional territories. A state of this nature is entirely different from the modern state. The looseness of its internal organization and the self-sufficient character of its regional divisions greatly magnified the importance and difficulty of the problem of regional relationship and the vital necessity of a Key Economic Area as a material basis of unity. It is a state which properly regards its public works for water-control as a weapon, and its policies are consciously or unconsciously guided by the immediate urgency, practically always present, of strengthening its Key Economic Area.

It is important to point out, however, that the concept of the Key Economic Area, specially formulated for the study of the historical processes of the period from 255 B.C. to A.D. 1842, becomes inapplicable to the enormously changed conditions in China since the country was opened to world trade and the influences of industrialism in the middle of the nineteenth century. With the building of railroads, the development of industry and trade, the growth of oversea commerce, the importance of public works for water-control as political weapons have greatly been reduced. Agricultural productivity has also lost its controlling significance as a measurement of political and military power.

The chief political problem is no longer merely a

question of regional domination between the various parts of China, but one of the dismemberment of China by the imperialist powers. The Treaty ports, serving as the base of economic and political operations of the powers, have grown into powerful economic and political centres of gravity, each port dominating a major section of China. The different ports, notably Shanghai, Hankow, Canton, and Tientsin, draw the economic and political life of the country in different directions and thus create a new situation of regional division and internecine struggle. The geographical outline of the new regional situation does indeed still coincide roughly with the old divisions, marked chiefly by topographical conditions ; but the economic basis and meaning of the new regionalization are entirely different from those of the old. If the term Key Economic Area is to be applied to the new situation at all, its meaning must be greatly stretched and practically redefined. The Yangtze valley, for instance, is no longer the Key Economic Area in the sense the term is used in this monograph, though it still seems to occupy a position of dominance in present-day China. The opening of China closed an epoch in its history and new concepts must be formulated to describe new relationships and analyse new conditions.

BIBLIOGRAPHY

(With Notes on Important Chinese Sources)

1. BIBLIOGRAPHICAL AND GENERAL REFERENCE

(a) In Chinese

Ch'ü Hsüan-ying, *Fang-chih K'ao Kao.* (*Preliminary Studies in Gazetteers.*) First Edition, 1930. Published privately. Only the first series (in 3 volumes) has been published. It is a descriptive bibliography of gazetteers which, when completed, will describe some 1,500 gazetteers in the private library of a Tientsin collector, Jen Feng-pao.

Fan Wen-lan, *Cheng-shih K'ao Liao.* (*Brief Studies of the Dynastic Histories.*) Published by the Cultural Society of Peiping, Peiping, China, 1931. 294 pages. A very handy and reliable volume for looking up facts about the twenty-five dynastic histories.

Mao Yung, *Chung-kuo Nun-shu mu-Lu Hui-p'ien.* (*Bibliography of Chinese Literature on Agriculture.*) Issued by the University Library, University of Nanking, 1924. This bibliography lists 286 items of literature on " Water-benefits."

Ou-yang Ying, *Chung-kuo Li-tai Chiang-yü Chan-cheng He-t'u.* (*Historical Atlas of China, illustrating historical administrative divisions and wars.*) Third Edition. Wuch'ang, Ya-Hsin Geographical Society, 1933. Consists of 46 maps with notes. Not very satisfactory, but the best on the subject.

Su Chia-yung, *Chung-kuo Ti-li Yen-ke T'u.* (*Historical Atlas of China.*) Third Revised Edition. Shanghai, Jih-hsin Geographical Society, 1930.

Tsang Li-he, et al., *Chung-kuo Ku-chin Ti-ming Ta Tz'u-tien.* (*A Comprehensive Dictionary of Ancient and Contemporary Chinese Geographical Names.*) Shanghai, Commercial Press, 1931. 1,410 pages. The best work on the subject. None of the other volumes on the same subject is as comprehensive.

Yüan T'ung-li, et al., *Kuo-li Pei-p'ing T'u-shu-kuan Fang-chih Mu-lu.* (*Bibliography of Gazetteers in the National Library of Peiping.*) Published by the National Library of Peiping, 1933. This bibliography covers the largest collection of its kind in the world, with over 5,200 titles, including many rare volumes.

(b) In Western Languages

Cordier, Henri, *Bibliotheca Sinica.* Dictionnaire bibliographique des ouvrages relatifs à l'Empire chinois . . . 2 ed., rev., cor. et considérablement augentee. . . . Paris, E. Cuilmete, 1904-08. 4 v. Supplément et Index, 1922-24. Paris, P. Couthner.

GOODRICH, L. C., and FENN, H. C., *A Syllabus of the History of Chinese Civilization and Culture.* Illustrated with maps and chart. New York, The China Society of America, 1929. 51 pages. This syllabus includes a detailed and well-selected bibliography of literature on Chinese history available in the English language.

HOANG, P. Concordance des Chronologies Néoméniques chinoise et européenne, Variétés Sinologiques, No. 29. Shanghai, Imprimerie de la Mission Catholique, 1910.

OXENHAM, E. L., *Historical Atlas of the Chinese Empire from the Earliest Times down to the Present Great Ching Dynasty*, giving the names of the chief towns and the metropolis of each of the chief dynasties of China. Second Edition, giving the original Chinese maps with their English counterparts. London, The Royal Geographical Society, 1898.

PLAYFAIR, G. W. H., *The Cities and Towns of China. A Geographical Dictionary.* Second Edition. Shanghai, Kelly and Walsh, 1910, xii + 582 pages.

TCHANG, MATHIAS. Synchronismes chinois, Chronologie complete et concordance avec l'ére chrétienne de toutes les dates concernant l'histoire de l'Extrême-Orient (Chine, Japon, Coreie, Annam, Mongolie, etc.) (2357 av. J.-C.–1904 apr. J.-C.). Variétés Sinologiques, No. 24. Chang-hai, Imprimerie de la Mission Catholique, 1905. The best reference for historical dates.

2. SPECIAL WORKS ON WATER-CONTROL IN CHINA

(a) IN CHINESE

CHIN PU, *Chih-ho Fang-liao.* (*Plans for Regulating Rivers, notably the Yellow River and the Grand Canal.*) 10 *chüan*. Edited by Tsiu Ying-chieh, 1767. The author was an outstanding Ch'ing official whose accomplishments were largely along the line of water-control. The book includes detailed maps of the Yellow River, the Grand Canal, the Hwai River, documentary material in connection with the author's work as the highest water-control official, a chronicle of flood and flood-control activities on the Yellow River, and researches into the historic route of the Yellow River, Grand Canal and other rivers, and on the conditions of lakes and ponds. At the end is appended an essay on the principles of water-control technique by Ch'en Hung, the eminent water-control expert who served as Chin Pu's secretary throughout the latter's career. This book ranks with P'an Chi-Hsün's *Ho-fang Ch'üan-shu* (*Handbook for River-control*) as a leading classic in the field and was regarded by later officials as a guide in their work.

FU TSE-HUNG, *Hsin-shui Chin-chien.* (*Golden Mirror of the Course of Rivers and Canals.*) 175 *chüan*, 1725. The *Imperial Catalogue* considers it the most detailed and comprehensive treatment of the subject of canalization for irrigation, transport and river-control. It consists mainly of extensive extracts from various primary sources. It contains many interesting maps of the various major streams. (*Imperial Catalogue*, *chüan* 69, p. 6.) In 1831, a supplementary work of 156 *chüan*, including one *chüan* of maps, was published under the editorship of P'an Hsi-en, assistant-Director-General of Rivers in Kiangnan, who completed the work begun by Li Shih-hsü, General in the Army and Director-General of Rivers in Kiangnan.

Hu Wei, *Yü-kung Chui-chih*. (*Notes on the Yü-kung*.) 20 *chüan*. First published in 1705. A very detailed and authoritative commentary on the *Yü-kung* or "*Tribute of Yü*." The field covered by the researches is so extensive that it can properly be regarded as a treatise on the historical geography of China. The researches into the change of the course of the Yellow River during historic times, and the one *chüan* of maps illustrating the changing course of the Yellow River during the Han, T'ang, Sung, Yüan, and Ming dynasties, are among its best features and form the basis of practically all later works on the subject. The *Imperial Catalogue* states that Hu Wei consulted almost all the existing commentaries on the *Yü-kung*, together with local gazetteers and maps, and it considers the book the best among "several tens" of commentaries on the *Yü-kung* written by scores of scholars during the Sung, Yüan and Ming dynasties. (*Imperial Catalogue*, *chüan* 12, p. 85.)

K'ang Chi-t'ien, *Ho-chü Chi-wen*. (*Notes on Rivers and Canals*.) 20 *chüan*. Hsia Yin-T'ang Edition, 1804. In an introduction to this book, the famous Tungchen scholar, Yao Nei, referred to the rich practical experience in water-control which the author gained while serving as Director-General of Rivers and Canals in Kiangnan. He was known for his earnestness in his work and frequently personally participated in field-work together with his subordinates. The book shows great learning, practical knowledge and historical insight. Practically all the major water-control activities during the various dynasties are discussed. One of the best books on the subject.

Ku Shih-lien, Shui-li Wu Lun. (*Five Essays on "Water-benefit."*) Published in 1655 as part of *Lu-Hsian-Tsi Ts'ung-shu*.

Liu Kwan-li, *Chung-kuo Tu-mu Hsin-cheng*. (*Engineering Policies in Chinese History*.) Peking, Publication Bureau of the Ministry of the Interior, 1919. 198 pages. The book is very badly written and shows that the author has no understanding of his subject, but it is useful as a collection of factual material.

P'an Chi-hsün, *Ho-fang Ch'üan-shu*. (*Handbook for River-control*.) 12 *chüan*. Publication date not clear, but the author's own preface was dated 1590. The author served as Director-General of the Yellow River and Grand Canal four times, totalling twenty-seven years of active service during the reign of Chia Ch'ing (1522-1566) and Wan Li (1573-1620). He was unquestionably the greatest authority on the subject during the Ming dynasty. The book contains a detailed map of the Yellow River and Grand Canal, with annotations and comments. It includes imperial edicts and memorials to the throne and other documents regarding river-control and water-transport, and illuminating discussions on various phases of the subject by the author, written in the form of polemics with an imaginary opponent who doubted his views. The book was, for centuries, taken as a guide by practical workers in the field. The *Imperial Catalogue* of the Ch'ing dynasty informs us that "although changes in methods were necessary to fit changing circumstances, yet experts in river-control always take this book as a standard guide." (*Imperial Catalogue, chüan* 69, p. 3.)

SHIH TU-CHEN, *Kiang Han Ti-Fang T'u K'ao.* (*Atlas and Notes on Dikes on the Yangtze and Han Rivers.*) 3 *chüan.* Published in 1568. The book contains a very interesting long introduction on the question of flood-control in the Yangtze valley. About two pages of notes are attached to each map in the series.

SUNG HSI-SHANG, *Shuo Hwai.* (*Discussion on the Hwai River.*) Nanking, Chin-hua Press, 1929. 152 pages. With maps and diagrams.

TUNG HSÜN-H'E, *Chiang-pei Yun-chen.* (*Handbook on the Course of the Grand Canal North of the Yangtze.*) 40 *chüan.* Published in the latter years of the Ch'ing dynasty, exact date uncertain. It contains a detailed map of the Grand Canal, names of places along its course, and distances between them. Practically all reservoirs, and the locks along the whole course of the canal, from Peiping to Yangchou, are recorded.

(*b*) IN JAPANESE AND ENGLISH

AOYAMA, SADAO, "Study on the Pien Canal in the Period of T'ang and Sung" (in Japanese). The *Toho Oakuho, Journal of Oriental Studies,* Tokyo, No. 2 (December 1931), pp. 1-49, with map. Published by the Academy of Oriental Culture. Tokyo Institute, Tokyo.

"The Grand Canal of China," by W. R. CARLES. *Jour. China Br. R. As. Soc., N.S.,* XXXI, No. 1, 1896–97, pp. 102-115.

"The Wonderful Canals of China," by F. A. KING. *The National Geographic Magazine,* Washington, Oct. 1912, pp. 931-958: maps and illustrations.

3. GAZETTEERS AND OTHER GEOGRAPHICAL LITERATURE

(a) Gazetteers, or Local Historical Geography in Chinese (arranged alphabetically according to names of provinces, prefectures, districts, etc.)

Ho Shao-chi, et al., *Anhwei T'ung-chih.* (*Gazetteer of Anhwei Province.*) 360 *chüan.* Printed in 1877. *Chüan* 61-64 on rivers and canals, *chüan* 65-68 on "water-benefits."

Shen Yi-chi, et al., *Che-chiang T'ung-chih.* (*Gazetteer of Chekiang Province.*) 283 *chüan.* Preface dated 1736. Printed in 1899. *Chüan* 52-61 on "water-benefits."

Huang P'eng-nien, et al., *Chi-pu T'ung-chih.* (*Gazetteer of Chip'u or the Metropolitan Area.*) 300 *chüan,* submitted to the throne in 1871. Printed in 1910. *Chüan* 75-91 on rivers and canals.

Chen Shou-chi, et al., *Fu-chien T'ung-chih.* (*Gazetteer of Fukien Province.*) 278 *chüan.* Completed 1829. Printed in 1868. *Chüan* 31-37 on "water-benefits."

Sun Hao, et al., *Ho-nan T'ung-chih.* (*Gazetteer of Honan Province.*) 80 *chüan.* Preface dated 1869. Printed in 1902. *Chüan* 12-16 on river-control, *chüan* 17-19 on "water-benefits."

Tseng Kuo-ch'üan, et al., *Hu-nan T'ung-chih.* (*Gazetteer of Hunan Province.*) 314 *chüan.* Preface dated 1885. *Chüan* 46-47 on dikes and dams.

Chang Chung-Hsin, et al., *Hu-pei T'ung-chih.* (*Gazetteer of Hupeh Province.*) 172 *chüan.* Printed in 1921. *Chüan* 39-42 on dikes and flood-control.

Li Ti, et al., *Kan-su T'ung-chih.* (*Gazetteer of Kansu.*) 50 *chüan.* Preface dated 1736. *Chüan* 15 on "water-benefits."

Huang Chih-chüng, et al., *Kiang-nan T'ung-chih.* (*Gazetteer of Kiangnan.*) 200 *chüan.* Preface dated 1736. *Chüan* 49-66 on rivers and canals. Kiangnan covered the present provinces of Kiangsu and Anhwei. No gazetteer of Kiangsu province has been published. Hence, for information on Kiangsu province, especially data for chapter 3, the material in this gazetteer relating to the present territory of Kiangsu has been used.

Liu Yi, et al., *Kiang-si T'ung-chih.* (*Gazetteer of Kiang-si Province.*) 185 *chüan.* Preface dated 1881. Printed 1882. *Chüan* 62-64 on "water-benefits."

Ching Tao-mu, et al., *Kuei-chou T'ung-chih.* (*Gazetteer of Kueichow Province.*) 46 *chüan.* Submitted in 1741. No significant data on water-control.

Hu Chien, et al., *Kuang-hsi T'ung-chih.* (*Gazetteer of Kwangsi Province.*) 279 *chüan.* Preface dated 1801. *Chüan* 117-120 on "water-benefits," but no chronological data.

Chen Ch'ang-chi, et al., *Kuang-tung T'ung-chih.* (*Gazetteer of Kwangtung Province.*) 334 *chüan.* Preface dated 1822. *Chüan* 115-119 on "water-benefits."

Wu Chung-Chi, et al., *Ling-Yin Hsien-chih.* (*Gazetteer of Ling-yin District,* Honan Province.) 8 *chüan.* 1660 Edition.

Wang Hsuan, et al., *Shan-si T'ung-chih.* (*Gazetteer of Shansi Province.*) 184 *chüan.* Printed in 1887. *Chüan* 66-69 on "water-benefits."

Yüeh Chun, et al., *Shan-tung T'ung-chih.* (*Gazetteer of Shantung Province.*) 36 *chüan.* Preface dated 1736. *Chüan* 18 on flood-control.

Shen Ch'ing-yai, et al., *Shensi T'ung-chih.* (*Gazetteer of Shensi Province.*) 100 *chüan.* Printed in 1735. *Chüan* 39-40 on "water-benefits." The best collected source material on historic water-control works in Kuanchung.

Yang Fan-tsan, et al., *Sze-ch'wan T'ung-chih.* (*Gazetteer of Szechwan Province.*) 204 *chüan.* Preface dated 1815. *Chüan* 23 on dikes and dams.

Fen Kuei-feng, et al., *Su-chou Fu-chih.* (*Gazetteer of Prefecture of Soochow.*) 150 *chüan.* 1881 Edition. *Chüan* 9-11 on "water-benefits." Contains very detailed documentary material. The famous "Letter on the Water-benefits of Su-chow," by Chia Tan, written in 1070, and "Letter on Water-benefits in Wu," by Tan E, written in 1088, both being well-known classics in the field, are reprinted here.

Wang Sung, et al., *Yün-nan T'ung-chih.* (*Gazetteer of Yünnan Province.*) 217 *chüan.* Printed in 1836. *Chüan* 52-54 on "water-benefits."

(*b*) Geographical Studies in Chinese

Chang Chi-chün (Chang Gee-yuen), *Pen-kuo Ti-li.* (*Geography of Our Country.*) Vols. 1 and 2. Shanghai, Commercial Press, vol. 1, 1926; vol 2, 1928. 268 and 490 pages. The best existing textbook in the Chinese language.

Ku Tsu-yü, *Tu Shih Fang-yü Chi-yao.* (*A Historian's Notes on Geography.*) 130 *chüan.* Edited by Pen Yuan-jin. Huang-tao T'ang Edition, 1774. A record of the geographical changes which have taken place in China from the earliest times down to the seventeenth century. It contains valuable discussions of the economic and strategic values of certain regions and places. The best book on the historical geography of China. An indispensable geographical guide to the study of Chinese history. A good index of the historical geographical names contained in this book, together with their present equivalents, was compiled by the Japanese scholar, Sadao Aoyama, who also contributed a learned biographical sketch of the author. This was published by the Academy of Oriental Culture (Tokyo Institute) under the title of *Shina Rekidai Chimei Yoran* in 1933.

KU YEN-WU, *T'ien-hsia Ch'ün-kuo Li Ping Shu.* (*A Work on the Geography of the Empire.*) 120 *chüan*. Completed in 1662. Lung Wan-yüeh Edition, published in 1811. This monumental work specially emphasizes the military aspects of Chinese geography, but also deals extensively with economic and political matters. The author was the first scholar of distinction to utilize the local gazetteers as sources on a large scale. He also quoted widely from memorials to the throne by officials of various dynasties and essays by outstanding scholars. An absolutely indispensable book for students of Chinese history, geography, economics, and political and military science.

Yü-kung, a fortnightly magazine on Chinese historical geography, published in Peiping under the joint editorship of Ku Chieh-Kang and Tang Chi-jang. Began publication in 1934.

(*c*) GEOGRAPHICAL STUDIES IN JAPANESE AND WESTERN LANGUAGES

BARBOUR, GEORGE B., "The Loess of China." *The China Journal of Science and Arts*, vol. iv, Nos. 8 and 9, Aug. and Sept. 1925. Pages 454-463 and 509-517.

BROOKS, C. E. P., *Climate Through the Ages: A Study of the Climatic Factors and their Variations.* London, Ernest Benn, Ltd., 1926. 424 pages.

CHU, CO-CHING, "Climatic Pulsation During Historic Times in China." *Geographical Review*, vol. xvi, New York, 1926. Pages 274-282.

CRESSEY, GEORGE BABCOCK, *China's Geographic Foundations: A Survey of the Land and Its People.* New York, McGraw Hill Book Company, 1934. 394 pages.

DOUGLAS, A. E., *Climatic Cycles and Tree-growth: A Study of the Annual Rings of Trees in Relation to Climate and Solar Activity.* Washington, D.C., Carnegie Institution, 1919, 123 pages. Vol. ii, 1928, 158 pages. Carnegie Institution of Washington, Publication No. 289.

FAIRGRIEVE, JAMES, *Geography and World Power.* London, University of London Press, 1917. 355 pages.

HANN, JULIUS, *Handbook of Climatology. Part I: General Climatology.* Translated by Robert de Courcy Ward. New York, The Macmillan Company, 1903. 429 pages.

HOSIE, ALEXANDER, "Droughts in China, A.D. 620-1643." *Journal of the North China Branch of the Royal Asiatic Society*, vol. xii (N.S.), 1878. Pages 51-89.

HOSIE, ALEXANDER, "Floods in China, 630-1630." *China Review*, vol. vii, 1878-79. Pages 371-372.

HUNTINGTON, ELLSWORTH, *The Character of Races*, as influenced by physical environment, natural selection and historical development. New York, Charles Scribner's Sons, 1924. 373 pages. Especially pages 148-204.

KENDREW, W. G., *Climate: A Treatise on the Principles of Weather and Climate.* Oxford, Clarendon Press, 1930. 320 pages.

KENDREW, W. G., *The Climates of the Continents.* Oxford, Clarendon Press, 1922. 374 pages.

LOWDERMILK, W. C., and SMITH, J. RUSSELL, "Notes on the Problem of Field Erosion." *Geographical Review*, vol. xvii, No. 2, New York, April 1927. 227 pages.

LYON, T. LYTTLETON, FIPPIN, ELMER O., and BUCKMAN, HARRY O., *Soils: Their Property and Management.* New York, The Macmillan Company, 1915.

MILLER, A. AUSTIN, *Climatology.* London, Methuen & Co., Ltd., 1931. 290 pages.

OMURA, KINICHI, *Political Geography of China* (in Japanese), 2 vols. Published by the Shanghai Oriental Society. Second Edition, 1916. An authority in the field. Contains very useful maps of water-courses in the various regions in China.

RICHARD, PÈRE L., *L. Richard's Comprehensive Geography of the Chinese Empire and Dependencies.* Translation by F. M. Kennelly, S.J. Shanghai, Tusewei Press, 1908.

RICHTHOFEN, FERDINAND PAUL WILHELM, FREIHERR VON, *Baron Richthofen's Letters*, 1870-72. Second Edition, Shanghai, "North China Herald Office," 1903.

RICHTHOFEN, F., Chinese Loess. *Geological Map*, May 1882, p. 293.

SION, JULES, *Asie des Moussons.* Paris, Armand Colin, 1928-1929. 2 vols.

TING, V. K., Professor Granet's "La Civilization Chinoise." *The Chinese Social and Political Science Review*, vol. xv, No. 2, July 1931, p. 268.

Wolfanger, Louis A., "Major World Soil Groups and Some of the Geographic Implications." *Geographical Review*, vol. xix, No. 1, New York, January 1929, pp. 106-107.

4. DYNASTIC HISTORIES AND OTHER WORKS ON CHINESE HISTORY

(For a description of the Dynastic Histories in the English Language see A. Wylie, *Notes on Chinese Literature*. Shanghai, Presbyterian Mission Press, 1922.)

(*a*) Dynastic Histories (arranged in Chronological Order)

Ssu-ma Ch'ien, *Shih Chi*. (*Historical Record*.) 130 *chüan*. Covers remote antiquity to 122 B.C.

Pan Ku, et al., *Ch'ien-han Shu*. (*Book of the Earlier Han*.) 120 *chüan*. Covers 206 B.C. to A.D. 24.

Fan Yeh, *Hou-han Shu*. (*Book of the Later Han*.) 120 *chüan*. Covers A.D. 25-220.

Chen Shou, *San-kuo Chih*. (*Record of the Three Kingdoms*.) 65 *chüan*. Covers 220-265.

Fang Chiao, et al., *Tsin Shu*. (*Book of the Tsin*.) 130 *chüan*. Covers 265-419.

Shen-Yüeh, *Sung Shu*. (*Book of the Sung*.) 100 *chüan*. Covers 420-478. (This is not the well-known Sung dynasty, but one of the minor kingdoms of the period of the Northern and Southern dynasties, 420-589.)

Hsiao Tzu-hsien, *Nan-ch'i Shu*. (*Book of the Southern Ch'i*.) 59 *chüan*. Covers 479-501.

Yao Szu-lien and Wei Cheng, *Liang Shu*. (*Book of the Liang*.) 56 *chüan*. Covers 502-556.

Yao Szu-lien, *Ch'en Shu*. (*Book of the Ch'en*.) 36 *chüan*. Covers 557-589.

Wei Shou, *Wei Shu*. (*Book of the Wei*.) 114 *chüan*. Covers 386-556.

Li Po Yao, *Pei-ch'i Shu*. (*Book of the Northern Chi*.) 50 *chüan*. Covers 550-589.

Lin-fu Te-feng, *Chou Shu*. (*Book of the Chou*.) 50 *chüan*. Covers 557-589.

Li Yen-shou, *Nan Shih*. (*Southern History*.) 80 *chüan*. Covers 420-589.

Li Yen-shou, *Pei Shih*. (*Northern History*.) 100 *chüan*. Covers 386-581.

WEI CHENG, et al., *Sui Shu.* (*Book of the Sui.*) 85 *chüan.* Covers 581-617.

LIU HSÜ, et al., *Chiu T'ang Shu.* (*Old Book of the T'ang.*) 200 *chüan.* Covers 618-906.

OU-YANGHSIU, SUNG CHI, et al., *Hsin T'ang Shu.* (*New Book of the T'ang.*) 255 *chüan.* Covers 618-906.

HSIEH CHÜ-CHENG, et al., *Chiu Wu-tai Shih.* (*Old History of the Five Dynasties.*) 150 *chüan.* Covers 907-959.

OU-YANG HSIU, *Hsin Wu-tai Shih.* (*New History of the Five Dynasties.*) 75 *chüan.* Covers 907-959.

T'o T'o, et al., *Sung Shih.* (*Sung History.*) 496 *chüan.* Covers 960-1279.

T'o T'o, et al., *Liao Shih.* (*Liao History.*) 116 *chüan.* Covers 907-1125.

T'o T'o, et al., *Kin Shih.* (*Kin History.*) 135 *chüan.* Covers 1115-1234.

SUNG LIEN, et al., *Yüan Shih.* (*Yüan History.*) 210 *chüan.* Covers 1206-1367.

K'E SHAO-MING, *Hsin Yüan Shih.* (*New Yüan History.*) 257 *chüan.* Covers 1280-1367.

CHANG TING-YÜ, et al., *Ming Shih.* (*Ming History.*) 332 *chüan.* Covers 1368-1643.

K'E SHAO-MING, et al., *Ch'ing Shih Kao.* (*Draft of Ch'ing History.*) 536 *chüan.* Covers 1644-1911. Officially banned in China because of its pro-Ch'ing (Manchu) tendencies.

(*b*) CHINESE HISTORICAL STUDIES IN CHINESE

Chung-kuo She-hui Shih Lun-chan. (*Discussions of the Social History of China.*) 4 vols. Collection of special contributions and articles originally published in the *T'u-shu Tsa-chih* (*Readers Magazine*). Publication date: vol. i, Third Edition, May 1932; vol ii, March 1932; vol. iii, not available so far; vol. iv, March 1933.

HSIEH WU-LIANG, *Chung-kuo Ku-tai T'ien-chih K'ao.* (*Researches in the Ancient Land System of China.*) Shanghai, Commercial Press, 1932. 95 pages. Analysis very superficial, but useful for its collection of source material.

HSÜ CHUNG-SHU, "On Some Agricultural Implements of the Ancient Chinese" (in Chinese). Academia Sinica, *Bulletin of the National Research Institute of History and Philology*, vol ii, part 1, pp. 11-59. Peiping, 1930.

KU CHIEH-KANG, *Ku Shih Pien.* (*Discussions of Ancient Chinese History.*) 2 vols. Peiping, Pu Society. Vol. i, 1926, 104/286 pages; vol. ii, 1930, 545 pages.

KU YEN-WU, *Lieh-tai Tse-ching Chi.* (*Records concerning Capitals during the Various Dynasties.*) 20 *chüan.* Pi-lin-lang Kuan Chun-shu. Published in 1907. First two *chüan* contain a general discussion of the selection of capitals, while the rest of the book is devoted to a consideration of the various seats of capital in historic times, such as Kuanchung, Loyang, Chengtu, Chiennieh, including detailed information on the year of the construction of the cities and palaces, temples, etc. The *Imperial Catalogue* commends the book for its detailed quotation from sources and careful research into the authenticity of the information, expressly recognizing the authority of Ku Yen-wu in geographical studies. (*Imperial Catalogue, chüan* 68, p. 15.)

KUO MO-JO, *Chung-Kuo Ku-tai Shih Nien-chiu.* (*Study of Ancient Chinese Society.*) Shanghai, The Modern Book Co., 1931. 6/313/20/29 pages.

LEE, MABEL PING-HUA, *The Economic History of China, with Special Reference to Agriculture.* New York, Columbia University Press, 1921. 461 pages.

LIU K'E-YÜ, *Chiu-t'ung T'ung.* (*The Nine Encyclopedias of Source Material on the Institutional History of China.*) 248 *chüan.* First published in 1902. This book practically reprints all the material contained in Tu Yu's *T'ung Tien,* Ma Tuan-lin's *Wen-hsien T'ung-k'ao,* and Cheng Ch'iao's *T'ung Chih,* and the two successive continuations of the three books by later authors. Only duplicate materials which originally appeared more than once in the various books have been eliminated.

LIU YI-CHENG, *Chung-kuo Wen-hua Shih.* (*History of Chinese Civilization.*) Vols. i and ii. Nanking, Chung-shan Bookshop, 1932. 528/544 pages. An interesting review of this book by Hu Shih appeared in the *Tsing Hua Journal* (in Chinese). Vol. viii, No. 2, Peiping, July 1933.

MIAO FENG-LING, *Chung-kuo T'ung-shih Kang-yao.* (*Outline of the General History of China.*) Vols. 1 and 2. Nanking, Chung-shan Bookshop, 1932. 418/400 pages. Not yet completed.

T'ANG CHI-JAN, "Chung-kuo Nei-ti Yi-min Shih—Hunan Pien" (History of Migrations within China—Essay on Hunan). *Shih Hsüeh Nien-pao.* (*History Year-book*), No. 4, June 1932. A very interesting article.

TAO HSI-SHENG, *Chung-kuo She-hui Shih-ti Feng-hsi.* (*A Historical Analysis of Chinese Society.*) Shanghai, New Life Book Co., 1929. 265 pages.

TAO HSI-SHENG, *Chung-kuo She-hui Yü Chung-kuo Kê-min.* (*Chinese Society and the Chinese Revolution.*) Shanghai, New Life Book Co., 1929. 320 pages.

TAO HSI-SHENG, *Hsi-han Ching-chi Shih.* (*The Economic History of the Earlier Han Dynasty.*) Shanghai, Commercial Press, 1931. 88 pages.

WAN KUO-TING (Wan Kwoh-Ting), *Chung-Kuo Tien-chih Shih.* (*An Agrarian History of China.*) Vol. i, Nanking, Nanking Bookshop, 1933. 394 pages.

WANG CHIH-HSIN, *Sung Yüan Ching-chi Shih.* (*The Economic History of the Sung and Yüan Dynasties.*) Shanghai, Commercial Press, 1931. 150 pages.

WANG TUNG-LING, *Chung-kuo Shih.* (Author's English title for the book: *History of the Various Dynasties.*) 4 vols. Peiping, Peiping Cultural Society, 1931. 384/800/590/430 pages.

YANG TUNG-CHUNG, *Pen-kuo Wen-hua-shih Ta-kang.* (*Outline of the Cultural History of Our Country.*) Shanghai, Peiping Bookshop, 1931. 542 pages.

(*c*) STUDIES IN CHINESE HISTORY IN JAPANESE AND WESTERN LANGUAGES

BISHOP, C. W., " The Rise of Civilization in China with Reference to its Geographical Aspects." *Geographical Review*, vol. xxii, No. 4, New York, October 1932, pp. 617-631.

CHEN ZEN, SOPHIA H., Editor. *Symposium on Chinese Culture.* Published by China Council of the Institute of Pacific Relations. Shanghai, China, 1931.

HUMMEL, ARTHUR W., " Ku Shih Pien " (Discussions of Ancient Chinese History), vol. i. *China Journal of Science and Arts*, vol. v, No. 5, November 1926.

HUMMEL, ARTHUR W., " What Chinese Historians are doing in Their Own History." *The American Historical Review*, vol. xxxiv, No. 4, July 1929.

LATOURETTE, KENNETH SCOTT, *The Chinese: Their History and Culture.* 2 vols. New York, The Macmillan Company, 1934. Has very detailed bibliography.

LI CHI, *Formation of the Chinese People: An Anthropological Inquiry.* Cambridge, Harvard University Press, 1928. 283 pages.

KOKIN, M., " *Ching-t'ien,*" *the Agrarian System of Ancient China.* Original in Russian. Chinese translation, by Ching Chi, bears the title, *Chung-kuo Ku-tai Shê-hui* (*Ancient Chinese Society*).

Shanghai, Li-ming Bookshop, 1933. With a long and valuable introduction by L. Madyar. Contains an exhaustive collection of source material on the subject, and very interesting discussion by the author.

MADYAR, L., *The Agricultural Economy of China.* (In Chinese.) Translated from the Russian by T. C. Chen and K. C. Peng. Shanghai (Originally published in Moscow in 1928. Chinese translation, 1930), 583 pages. A very illuminating scientific study.

NAGANO, AKIRA, *Studies of the Chinese Land System.* Original in Japanese. Chinese translation, by Lu Po, bears title, *Chung-Kuo T'u-ti Nien-chiu.* Shanghai, Hsin-sen-ming Bookshop, 1933. 406 pages.

RADEK, KARL, *Theoretica Analysis of Chinese History.* Translated from Russian into Chinese by Ke Jen under the title, *Chung-kuo Li-shih Li-lun ti Nien-chiu.* Shanghai, Hsin-k'en Bookshop, 1932. 256 pages.

SAFAROV, *History of the Development of Chinese Society.* Original in Russian. Chinese translation by Li Li-jen under the title, *Chung-kuo She-hui Fa-chan Shih.* Shanghai, Hsin-shen-ming Bookshop, 1932. 560 pages.

WITTFOGEL, KARL AUGUST, *Wirtschaft und Gesellschaft Chinas; varsuch der wissenschaftlichen analyse einer grossen asiatischen agrar-gescellschaft.* Leipzig, C. L. Hirschfeld, 1931.

5. MISCELLANEOUS REFERENCES

(a) IN CHINESE

LIANG CH'I-CH'AO, *Yin-pin-shih Wen-chi.* (*Yin-pin-shih Collecte Essays.*) Shanghai, Chung-hua Book Co., 1926 Edition.

WANG CHENG, *Nun Shu.* (*Book of Agriculture.*) 10 *chi* (chapters). A Yüan book. Date of first publication, 1314. Fu-yung-lo Edition, 1617. Very interesting book on the technical processes of agriculture, with illustrations.

(b) IN WESTERN LANGUAGES

BUCK, JOHN LOSSING, *Chinese Farm Economy.* Shanghai, Commercial Press, 1930. (Published for the University of Nanking and the China Council of the Institute of Pacific Relations.)

BUCKLEY, ROBERT BURTON, *Irrigation Works in India and Egypt.* London, E. S. F. N. Spon, 125 Strand, 1893. 348 pages.

KING, FRANKLIN HIRAM, *Farmers of Forty Centuries, or Permanent Agriculture in China, Korea and Japan.* Madison, Wisc., Mrs F. H. King, 1911. Illustrated.

MALLORY, W. H., *China, Land of Famine.* New York, Am. Geo. Society, 1926. 151 pages.

MARX, KARL, "The British Rule in India." *New York Daily Tribune*, 25th June 1853, p. 5.

METCHNIKOFF, LEON, *Les Grands Fleuves Historiques.* Paris, Hachette, 1889, in 12, 369 pages, cartes. Chapter XI (pp. 320-363); Houng Ho et Yangtze Kiang.

SIMKHOVITCH, VLADIMIR G., *Toward the Understanding of Jesus and other Historical Essays.* New York, The Macmillan Company, 1921. 165 pages. Note Essay on "Rome's Fall Reconsidered."

SMITH, A. DAIRL., *Italian Irrigation: A Report on the Agricultural Canals of Piedmont and Lombardy, addressed to the Honourable the Court of Directors of the East India Company.* Vol. i, Historical and Descriptive, and vol ii, Practical. London, W. H. Allen & Co., 1852. 434 pages.

VERNON-HARCOURT, LEVESON FRANCIS, *Rivers and Canals: The Flow, Control and Development of Rivers and the Design, Construction and Development of Canals, both for Navigation and Irrigation.* 2 vols. Oxford, Clarendon Press, 1896. 705 pages.

WEBER, MAX, *General Economic History.* Translated by Frank H. Knight. London, George Allen & Unwin Ltd., 1927. 401 pages.

INDEX

(Since the table of contents is sufficiently detailed to serve the purpose of a subject-index, an index of proper names only is given here.)

Printed by Turnbull & Spears, Edinburgh

图书在版编目(CIP)数据

中国历史上的基本经济区＝Key economic areas in Chinese history:英文/冀朝鼎著.—北京:商务印书馆,2014(2020.2重印)
(中华现代学术名著丛书.英文本)
ISBN 978-7-100-10501-9

Ⅰ.①中… Ⅱ.①冀… Ⅲ.①经济区—经济史—中国—古代—英文②水利经济—经济史—中国—古代—英文 Ⅳ.①F129.2

中国版本图书馆CIP数据核字(2013)第289475号

中华现代学术名著丛书
中国历史上的基本经济区
(英文本)
冀朝鼎 著

商 务 印 书 馆 出 版
(北京王府井大街36号 邮政编码100710)
商 务 印 书 馆 发 行
北 京 通 州 皇 家 印 刷 厂 印 刷
ISBN 978-7-100-10501-9

2014年9月第1版 开本710×1000 1/16
2020年2月北京第2次印刷 印张13 插页1

定价:39.00元